D1598443

COMPUTER-AIDED LEAN MANAGEMENT FOR THE ENERGY INDUSTRY

COMPUTER-AIDED LEAN MANAGEMENT FOR THE ENERGY INDUSTRY

Roger N. Anderson,
Albert Boulanger,
John A. Johnson,
and Arthur Kressner

Disclaimer: The recommendations, advice, descriptions, and the methods in this book are presented solely for educational purposes. The author and publisher assume no liability whatsoever for any loss or damage that results from the use of any of the material in this book. Use of the material in this book is solely at the risk of the user.

Copyright © 2008 by
PennWell Corporation
1421 South Sheridan Road
Tulsa, Oklahoma 74112-6600 USA

800.752.9764
+1.918.831.9421
sales@pennwell.com
www.pennwellbooks.com
www.pennwell.com

Marketing Manager: Julie Simmons
National Account Executive: Barbara McGee

Director: Mary McGee
Managing Editor: Stephen Hill
Production Manager: Sheila Brock
Production Editor: Tony Quinn
Cover Designer: Alan McCuller
Book Layout: Lori Duncan

Library of Congress Cataloging-in-Publication Data

Computer-aided lean management for the energy industry / Roger N. Anderson ... [et al.].
p. cm.
Includes bibliographical references and index.
ISBN 978-1-59370-157-4
1. Energy industries--Cost control--Data processing. 2. Energy industries--Labor productivity--Data processing. 3. Energy industries--Management--Data processing. 4. Industries--Energy conservation--Data processing. 5. Business logistics--Data processing. 6. Decision support systems. 7. Computer-aided engineering. I. Anderson, Roger N. (Roger Neeson), 1947-
HD9502.A2C645 2008
333.790285--dc22

2008011803

All rights reserved. No part of this book may be reproduced, stored in a retrieval system, or transcribed in any form or by any means, electronic or mechanical, including photocopying and recording, without the prior written permission of the publisher.

Printed in the United States of America

1 2 3 4 5 12 11 10 09 08

We dedicate this book to our families:

From Roger to Honor, Roger, Jon, and Forrest;

From Albert to my father Arthur, who instilled curiosity and a desire to improve how things work;

From John to Andrea and Allison;

From Arthur to Millie, David, and Diana.

In addition, this book is also dedicated to the perceptions of Robert Broadwater, Roice Nelson, and Rick Smalley. May their visions of affordable, abundant, secure and intelligent energy supplies for Planet Earth soon come true.

CONTENTS

PREFACE

Data in the energy industry are plentiful and generally of high quality, but the integration of those data into information and knowledge are the challenge. Because the data never get into best-practices form, the industry is plagued by never-ending one-offs in designs, builds, and operations that repeat mistakes over and over again. For example, the information challenge that energy companies face is illustrated by how well a company responds after a bad event. Energy companies can always reconstruct how that event evolved and break down specific elements that went wrong into often millisecond resolution. Sufficient data were clearly there in the first place. The challenge is, why can't we do that on a continuing basis—and predict these events before they happen? In this book, we discuss how to apply current and emerging lean technologies and methodologies that can create business capabilities to transform the energy companies' use of information and knowledge. Today, in industry and university research departments, we are evolving and developing tools and methodologies that we call computer-aided lean management (CALM): physical and work process coupling with machine learning software that, when combined into one integrated system model of the business, provides dramatic improvements throughout the enterprise.

As we struggle to sustain and advance our modern society, we look to energy and alternative energy companies to help us in providing comfort, food, mobility, and security in our lives. As we look to the future of energy availability, our present way of extracting fossil fuels and converting and delivering this energy to society is not sustainable. We write this book to provide a road map for the energy industry—oil and gas, as well as electric utilities—to guide the transformation of the way they do business today to meet their growth imperative given the uncertainties of the future world. We extend into the energy industry the best practices available in computational sciences and lean management principles that are presently being used by the aerospace and automotive industries. We describe how to create lean business capabilities for field-oriented, asset-heavy infrastructure companies. These technologies and methodologies can be used to continuously lower costs, while improving quality and service. We also discuss potential future developments in computational sciences on the horizon that will provide these infrastructure businesses with new options to create truly smart businesses, thus enabling them to develop and deploy the intelligent grid that will be required for the electric economy that is coming.

We hope this book provides you with a view on how to transform your business. Imagination and ability require you to first visualize what is possible. We hope we have helped with that. We wish to thank the many friends and colleagues who supported this effort: Fred Seibel of CALM Energy, Dave Waltz, Marta Arias, and Phil Gross of the Center for Computational Sciences at Columbia University; Maggie Chow, Serena Lee, and the engineers of the Edison Program at Con Edison; and Sally Odland of the Lamont-Doherty Earth Observatory. We also thank our wonderful editorial associates, John D. Roy, Professor Honor O'Malley of Columbia University, and Andrea Chebuske of CALM Energy.

1 INTRODUCTION

Management of oil, gas, electricity, renewable, and other energy businesses requires massive investments in interdependent, heavyweight infrastructure, as well as simultaneous attention to disparate market forces. These businesses are dominated by risk from uncertainties such as weather, market variations, transportation disruptions, government actions, logistics difficulties, geology, and asset reliability. We predict that in the future, these uncertainties and accompanying business inefficiencies can be continuously acted upon to increase profitability through the use of computational decision-making tools that will enable rich new opportunities for additional value creation.

Energy businesses routinely make strategic decisions to sell, buy, build, use, and reuse infrastructures of all kinds. Costly decisions about the maintenance and redesign of these facilities are continuously made. Assets have to be modernized and modified to provide reliable and continuing value to the customer, or they risk being made obsolete by changing technologies, regulations, and/or market forces.

In infrastructure businesses that manage field assets, such as the energy industry, uncertainty is the prime impediment to profitability, rather than the maintenance of efficient supply chains or the management of factory assembly lines. Energy products must be delivered 24/7—and often to a global market. Bad events related to weather, equipment failures, industrial accidents, government regulations, and train wrecks (both real and metaphorical) must be managed. In this book, we present a vision of the development of a methodology within a company that will ultimately steer efficiently through these uncertain seas. We present both the present state of transformational business capabilities and the technologies that are being worked on today. They will hopefully extend the transformation of business capabilities to full implementation of *computer-aided lean management* (CALM) in the near future. We are specifically addressing the energy businesses that involve the management of field infrastructure, but the principles and techniques will apply generally to all asset-intensive industries. CALM is an extension of lean management that includes rigorous, enterprise-wide, computer-aided decision support, modeling, simulation, and optimization. Most of the components of this system exist today, but scaling to the enterprise level is likely to be a 10-year project. Throughout the book, we will be careful to identify those tools and techniques that are working in energy companies today and those that are still under development and testing for future deployment.

Computer assistance is critical to the management of the ever-increasing volume and scope of information that is guiding decision-making. Today, companies experience crippling difficulties in efficiently collecting, handling, comprehending, and communicating that information. CALM will enable its efficient capture, cleaning, and use to enable optimal management of assets and people in diverse locations. This will require the informed, dynamic planning and coordination that CALM seeks to make possible.

In too many energy companies, information is not coordinated. Often, separate spreadsheets of versions of the same information are "owned and guarded" in multiple computer systems and many different databases. Consequently, real-time coordination, scheduling, planning, and execution are impossible without an army of planners and analysts.

An understanding of how energy systems are working in real time is equally difficult today. Following a major incident, every possible piece of information is gathered and investigated for months on end. Only then can management understand exactly what happened, when and where the incident started and ended, and what and how external drivers contributed. Although clarity is hard to achieve, energy companies can invariably document the root causes of big, bad events. This understanding is arrived at through analysis of the data that already existed and were routinely collected daily. The inevitable question is, why couldn't we have seen this coming beforehand? In the future, proper information management and its interpretation, coupled with computational assistance, will make uncertainties visible so that CALM companies can see what is coming.

Transformational improvements in the management of costs, resiliency, sustainability, and cycle times will emerge during the execution of CALM because of its software rigor and reproducibility. Failure models will become predictive, costs will be analyzed for the life of the asset, risk and return will be estimated for all investments, and solutions will be identified quickly and scored for their effectiveness. Metrics that measure whether the work is being done on schedule and on cost will be fed back into a computer-learning system so that only the most efficient processes become the best practices.

CALM will ensure that accurate data are getting to the right people at the right times, so that the performance of all critical systems is improved over their lifetimes.

The CALM process of continuous improvement can be efficiently taught, executed, and ironically, improved continuously. This book presents the steps necessary to recognize and manage the risks while capturing the rewards that come from optimizing responses to the multitude of uncertainties inherent in field asset businesses, now and into the future. For example, CALM will enable operational innovation through the deployment of flexible software that can be customized, resulting in continuous refinement of the ways a company does business. CALM software implementation methodologies reduce operating risk, enhance customer service, and increase reliability by putting the development in the hands of the user all along the way. Information technology (IT) departments and contractors will work with the ultimate user to ensure that that any new software development for computer aiding will be doing what it is designed to do and that it will be used.

Continuous improvement and innovation will be enabled through the creation of one integrated system model (ISM) of the business. The ISM will coordinate deployment of CALM methods and techniques. The ISM will merge three major software systems that are not yet standard to field-intensive industries like the energy business:

- *A plant model (PM).* A high-resolution model of the entire physical and business infrastructure that is to be managed.
- *A business process model (BPM).* A dynamic and detailed process and workflow model that will track information and measure performance on a daily basis with the goal of optimizing all business processes.
- *Computational Machine learning (ML).* A modern computer analysis methodology that uses historical and real-time operational data to predict, prioritize, and optimize critical processes of a company—including optimization of the business itself.

CALM will provide a commonsense approach for running the business, by measuring the results of actions taken and using those measurements to design improved processes in order to drive out inefficiencies. This book outlines the strategic and tactical framework that will be necessary to create a CALM energy company of the future. In addition, we describe in detail the dynamic tools and techniques that are available today and the technologies and mathematical improvements that are under development for future deployment. In particular, CALM is driving the establishment of feedback loops so that intelligent computational systems can learn how best to take actions that produce continuously improving performance. The closing of all critical feedback loops for a large enterprise is probably still 10 years out, but the testing and development is under way.

The ISM will provide the tools needed to visualize the competitive landscape and working environment of the company. The ISM will provide forward-looking simulations of scenarios and evaluation of real options, so that alternatives can be explored to find the innovations that will most likely optimize the company's business performance. A company will need these tools in order to become more adaptive and better able to perform successfully in the future, as the business changes under foot.

The adoption of CALM will provide any enterprise with a significant competitive advantage, whatever its existing core business. Consequently, the company will be able not only to deliver better internal quality control and incentives to produce products and services for its customers but also to identify and capture the benefits from external opportunities such as mergers and acquisitions. Significant post-merger synergies can be recognized through quick integration of products and operations using these newly developing methods and tools. The stockholders, customers, and employees will all win.

MISSION

The CALM methods and technologies in this book are focused on future improvement for *field industries* that conduct daily work in the urban and rural outdoors, rather than inside factory floors and enclosed assembly lines where lean principles are well established. Field-intensive industries, such as energy businesses, are different from those centered on the factory floor. While striving to lower capital, operations, and maintenance costs, they also wish to increase the profitability of their long-life field assets. There are enormous business risks inherent in the operation of such long-life assets: They can quickly degrade, with potentially catastrophic consequences. In addition, we seek to speed the transition to CALM in countries—such as China, India, and Russia—that are the new business environments of the 21st century and are therefore less inhibited by the need to transition from legacy systems.

Figure 1–1 illustrates the hoped-for future results from deployment of CALM. The bottom represents costs to run the business from a unit-of-supply perspective, and the top depicts profitability from a deployed-asset standpoint. As can be seen, a significant reduction will be possible in the operational, maintenance, and capital costs needed to create and sustain the business.

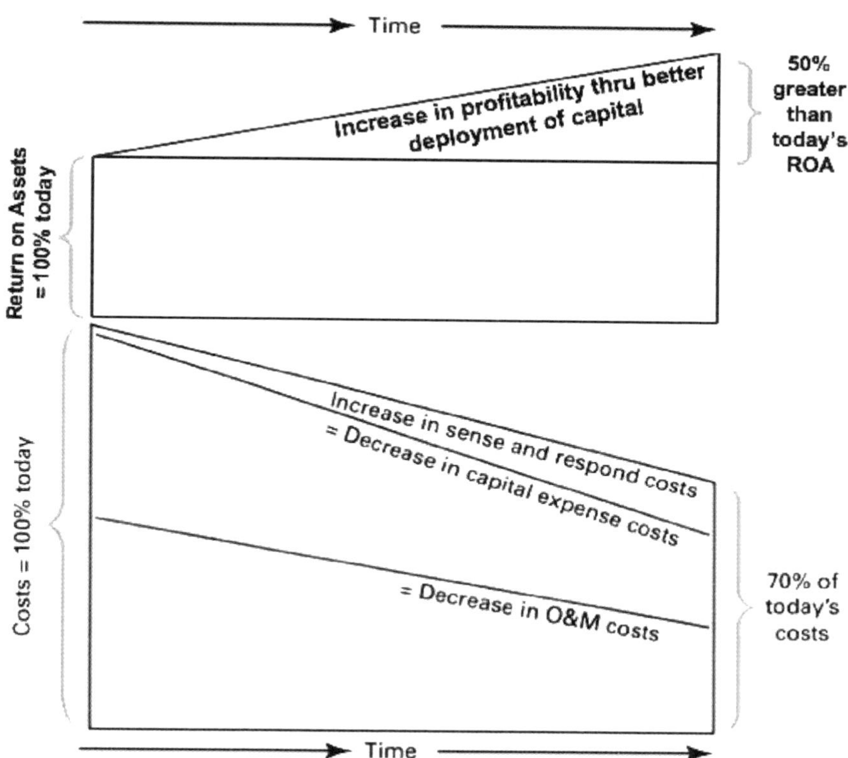

Fig. 1–1. Typical deployment resulting from lean implementations across many industries. Costs are decreased by up to 30%, and returns on assets are increased by up to 50%.

CALM will require a modest cost increase for what we call *sense-and-respond* technologies. The uses of two-way communications with more sensors in the field and significantly more computation will enable, in turn, more efficient operations. A conservative 30% reduction in overall costs has been consistently achieved in lean implementations in other industries.[1] There is also a track record for dramatic improvement in profitability through better capital deployment. The company will be able to realize more profit with fewer assets by using CALM methodologies.

In regulated businesses, CALM will enable the firm to exceed expected performance while recognizing and reducing true business risks, thus consistently meeting or exceeding the agreed-to return-on-investment (ROI) targets. It has been the experience of most companies that undergo lean implementations that cost and value can be extracted quickly by improving the management of "low-hanging-fruit"—obvious assets that were not previously managed efficiently. Significant growth opportunities and increased customer value can come from the reduction of inefficiencies that have been identified by traditional lean analysis techniques.

METHODOLOGY

The CALM methodology begins with the existing supervisory control and data acquisition (SCADA) and IT systems and software of a company. Critical inputs are integrated into an ISM encompassing customers and business processes, so that accurate simulations can be computed for better decision-making. Then, the simulations are visualized and optimized so that performance can be improved for customers, shareholders, and employees—simultaneously. A fly-by-wire business process operates via computer control, with little human intervention, to manage the process if all is going well. Management can focus on early identification and correction of what is not going so well. A flexible, always-in-the-money decision system will be created in the future in the form of a branching tree of real options, similar to forks in the road. The decision system always provides at least two—and often more—alternative actions to be evaluated for present and future value. The most profitable real-option choice will be recommended and risks identified. (See fig. 1–2.)

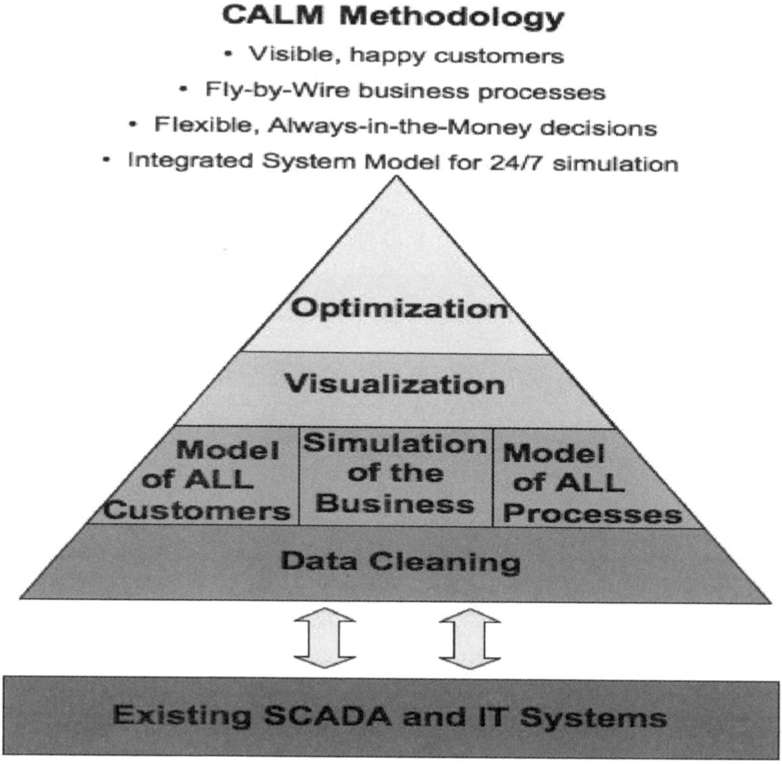

Fig. 1–2. The CALM methodology. Take existing SCADA and IT software systems; clean the data (always); build an ISM of customers and processes, so that simulations of the business can be run 24/7; and visualize the results, so that optimization can result in better performance for customers, shareholders, and employees—a win-for-all strategy.

CALM implementation uses a formal analysis methodology focused on how well technology, processes, and structures are being integrated within projects in a company today. Additional business capabilities are identified that need to be developed within the company to implement CALM fully (fig. 1–3).

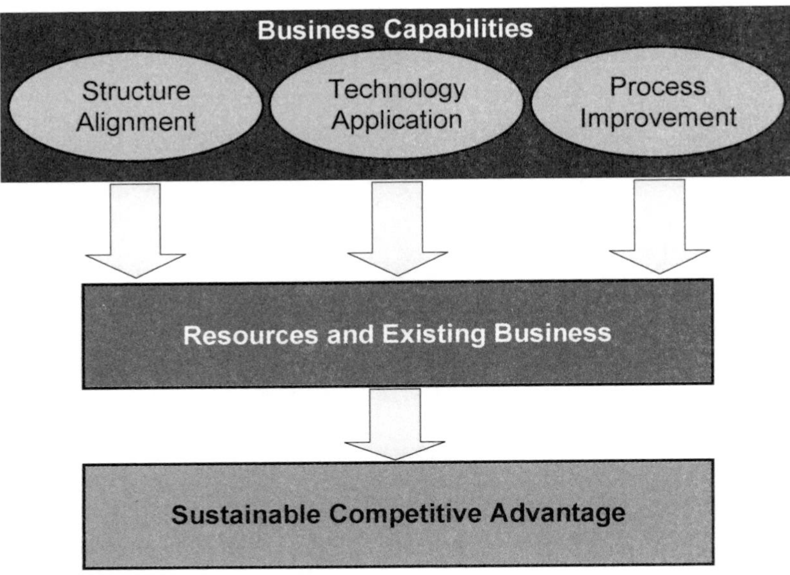

Fig. 1–3. Improvement of business capabilities. CALM enables technology applications, streamlining of processes, and a lean organizational structure. This, in turn, enables the company to invest in and improve the business in a sustainable way.

Entrenched past practices of many infrastructure-based, risk-averse industries can make CALM implementation difficult. Employees must be focused on the same goals through the structural realignment of organizations, interrelationships must be managed, and well-planned and well-thought-out incentives provided. CALM asks that plant, process, and economic models be constructed that integrate strategy into tactical execution to improve operational goals. A structured governance mechanism should be used to monitor and evaluate success toward the attainment of these goals. This approach will create a coherent business structure with an awareness of what to focus on and what to avoid, so that everyone will be on the same page and working to create enterprise value.

MOON SHOT

Going to the moon took intensive planning, scheduling, development of completely new systems, the invention of new technologies, many iterations of discovery and testing, and integration of multidisciplinary teams from many agencies and organizations. The same holds true, although on a smaller scale, for CALM implementations in the energy industry; coordination needs to be planned, scheduled, budgeted, and sustained with sufficient resources to execute effectively throughout the company.

Investments in improvement normally yield returns within the first year. The low-hanging fruit—such as high-value software implementation projects—can be quickly focused on the customer, whether internal or external, to yield immediate value. Contrast that with traditional, multiyear IT software implementation programs that create enormous risk to the enterprise and consistently fail to meet the requirements of employees and customers alike, to say nothing of the usual cost overruns. In CALM implementations, alpha replaces beta as the first version of a software innovation that employees and customers see and use. Immediate payback in both usefulness and efficiency of the software consistently results (fig. 1–4).

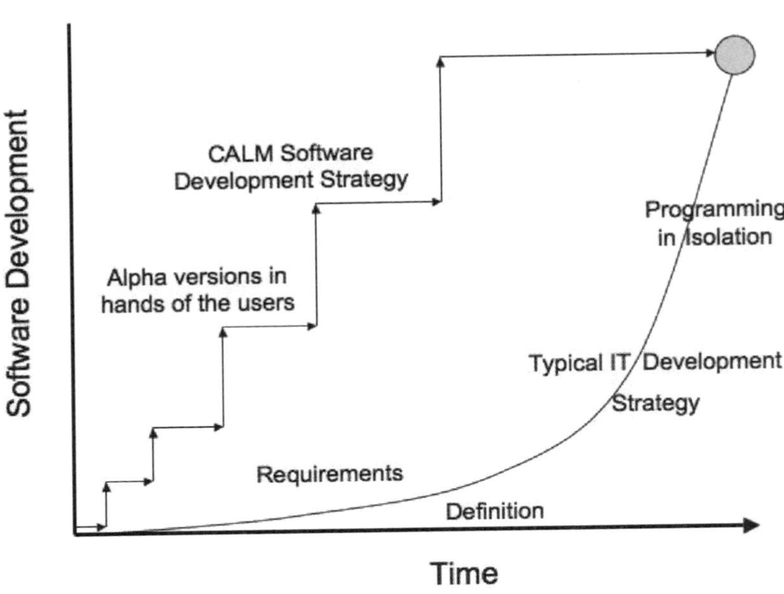

Fig. 1–4. The CALM software development strategy. Alpha versions are supplied to users as soon as possible so that multiple revisions and additions can be made; thus, when version 1.0 of the software is issued, it represents what the end user really needs, wants, and will use. In addition, less time will be required in order to develop the first usable software.

Enabling an energy company to implement CALM requires a combination of best practices, including systems engineering methods that are foreign to most in the energy industry. Systems engineering is used to build complex projects, such as aerospace vehicles (e.g., Skunk Works at Lockheed Martin, PhantomWorks at Boeing) and advanced automobiles (e.g., the Lexus that parks itself). We present templates for creation of an integrated master plan (IMP) and an integrated master schedule (IMS), which are used in systems engineering projects to manage all personnel, processes, and resources; these tools provide critical computer aids so that long-range improvement goals are achieved—on budget and on schedule.

The IMP and the IMS are live documents that are continually updated and refined. CALM projects are planned and scheduled on the computer before work begins. Key initial requirements are the establishment of integrated process teams (IPTs) in business intelligence, BPM, ISM, and decision aids that are in charge of setting key performance indicator and ROI targets that are continually monitored and scored.

These IPTs contain one member each for all organizational silos being crossed in the implementation. They must be empowered by top management with responsibility, authority, and accountability (RAA).

Thus, the implementation road map begins at the top of the corporation. CALM presents a significant challenge to the rules of the game within a company, and only with executive support will it succeed in significant modification of the behavior inside the organization. This change will be brought about not by words and meetings, but by computer systems, best practices, and integrated processes that are enabled by rigorous software enforcement.

There are many agents-against-change entrenched in any organization, especially, it has been observed, if a company has a centralized IT department. It is human for some people in any organization to resist change. CALM changes decision-making authority, information availability, performance metrics, and implicit and explicit rewards. Some people just won't like that. For example, the performance metrics for IPTs focus on measurable value to the company, instead of on subjective or even measured performance related to the achievement of personal goals within a specific project.

The extensive use of data and simulations to run the company and to score performance is a game-changer. What, then, would be the migration path to incorporate these advanced tools into the ongoing operations of your company? As they say at Boeing, "How do we redesign the plane while it's in the air?"

In other industries, a common lean goal is the delivery of products and services in half the time at half the cost. In energy and other field-intensive asset industries, costs are not the be-all and end-all of profitability. Risk-adjusted opportunities for investment and better utilization of deployed field assets are lean goals as well. The benefits from creating and exercising the real options are enormous.

STRATEGIC GUIDE

In this section, we outline a strategic guide for a CALM implementation that an energy company may wish to consider incorporating into their strategic plan. The ultimate strategy is to become a performance leader by developing and deploying lean innovations and solutions that increase shareholder value, satisfy customer needs, and improve the quality of life in the shared community. CALM strives to create wins for all, including customers, employees, and shareholders. Regulated industries, such as electric and gas utilities, have the additional requirement to improve return at reduced risk by providing customers with reliable products and services at fair costs.

To execute the CALM strategy means to transform the internal operations of the business by lowering risk, driving out inefficiency, and enhancing asset reliability. This improves the customer experience and enables compliance with all the environmental and safety requirements and goals of the community. Implementation creates a sustainable corporate advantage through better coordination of company resources and exploitation of economies of scale and scope.

The key principles of CALM consist of setting up a lean management structure and associated policies, building a foundation for continuous improvement through a healthy reliance on computational efficiency, and communicating all needed information openly. This starts with requirements for high-quality data and information about how the company works and ends with development of an ISM of the entire business entity. Much of the data already exist within a company but are rarely used for decision-making purposes, except perhaps for incident investigations or management-initiated special studies. For these and new data sources to be used properly, additional principles have to be incorporated into the firm's objectives to make this information useful for the enterprise. Following are the more demanding principles for field industries that are usually not covered in books on lean management for manufacturing:

- *Enter data once and only once.* When data are entered accurately and made available to anyone or any system, transaction costs associated with dealing with the information are greatly reduced. Imagine an asset that from a data standpoint exists uniquely, in one relational database. All of its data are readily available for anyone to access over the life of the asset.
- *Capture accurate data in a timely fashion at the source.* We all remember the game of telephone tag, in which a verbal message was sent around a group but, by the time it returned to the originator, the content was not recognizable. The same holds true for capturing accurate data at the source. When more than one individual gets between data and transfers, errors pop up and inaccurate information will be proliferated. If a mechanic fills out a form three hours after completing a job, how accurate will the data be? When a clerk enters that information into a database three weeks later, how accurate will it be? Any uncertainty about an entry can cause hours of investigation or, as happens more often, erroneous entries.
- *Get data in a usable and accurate form.* Data quality is critical to comprehending the state of any system, and accurate data have proved their value over and over. This point should be made clear to all people who are charged with data entry. What's more, if a technician has an option to enter information in free form in any data field, that information will be difficult

to use in later analyses. CALM requires *poka-yoke* methods (simple templates designed to prevent free-form mistakes). They provide a consistent form of data entry so that quality is ensured. This normally requires intensively collaborative software design and many iterations between IT personnel and the users of these data fields until the software meets all technical needs without the potential for data entry substitutions. In addition, communications about how and why this information is valuable provides both IT developer and data entry technicians with an understanding of why their tasks are so critical. This ensures more accurate data and enables process innovation from technicians within lean organizations.

- *Develop efficient historical repositories, or data warehouses, for future use by anyone needing access to the data.* Making historical data available to everyone will be one of the first steps to enabling lean management through measurement of performance over time. Accurate time stamps and version controls are the most important entries that accompany any data stored in a modern data warehouse.
- *Innovations in software are encouraged by anyone in the organization.* Innovation by those directly involved in the processes can elicit a swarm of operational discoveries that, if properly managed, can create transformational changes from the bottom up. New one-off software is not needed to drive this innovation. CALM encourages the use of commercial off-the-shelf (COTS) software.
- *Develop user-defined applications.* Software designers are not the end-users, and they consistently prove this with the creation of poorly designed software that does not meet the requirements for the actual users. However, the end-users may not know what they want until they see it, and even then, they may realize that they want and need something else. The users of the software must have "skin in the game" to get the business requirements transformed into software solutions that they

will use. Also, if the users are granted ownership in defining the applications, they will be more inclined to manage the successful deployment of the software.

- *Make all applications available for continuous modification and improvement by the users.* Having the actual users responsible for continuous improvement sounds challenging, but in most cases, they are the only ones who understand the business objectives. In many cases, the users want to improve the ways in which they work but have been immobilized by rigid software, which requires that they perform tasks in only one way. Having flexible software that can be easily modified by the user with only modest help from IT developers will empower the end-user.
- *Develop business applications using an iterative lean software development methodology.* Putting alpha versions in the hands of the users as soon as possible allows them to realize what they do and don't want. The software can be cheaply morphed into a more valuable form before too many features are locked in. For example, software often stagnates in the development laboratory for months, spoiling like old fruit, while IT groups or outside consultants decipher functional specifications that took months to develop. When completed, the software may be so far off the mark of the user's requirements that it simply will not be used. This is the norm for IT groups and consulting firms that are not embedded as teams within the organization; they can't collaborate sufficiently to make speedy, iterative revisions to the software. Measuring how much new software is successfully used becomes a critical lean metric.
- *Drive toward the elimination of free-form uses of e-mail and spreadsheets.* Distribution of one-off pieces of information such as e-mailed analyses of spreadsheets results in information siloing, with no one outside the mailing list able to leverage or reuse what often are valuable analysis templates. In addition, more than half of the e-mails received by any manager will be a waste of time to read. For the sender, adding 20 names to a

carbon copy (CC) list provides cover and lets everyone know that you are "working." However, it wastes time and effort almost as much as going to too many meetings.

- *Properly measure the effectiveness of work done on assets to determine if it was successful or detrimental.* How many companies measure whether maintenance actions performed on assets actually fixed the problem? Without measuring the effectiveness of the maintenance, including asset reconfigurations of any kind, you will never know whether the work was helpful to the enterprise.
- *Create resiliency in customer usage through software intelligence.* Enabling elasticity in your customer's usage of your products creates the opportunity to reduce supply system redundancies that were needed to guarantee service at all costs.
- *Create distributed intelligence in the organization and systems.* Distributing intelligence and decision-making capabilities throughout the organization empowers people at every level and will be critically important to enabling operational innovation. Improved utilization of assets and human resources will result while enhancing reliability of the products. Who would have initially thought that great value can come from the development of free, open-source software? Consider the success of projects like Wikipedia. Prestigious *Nature* magazine recently found that "Wikipedia comes close to *Encyclopaedia Britannica* in terms of the accuracy of its science entries!"[2]
- *Create one ISM that can eventually run the entire company.* Through the creation of the ISM, data and information can be efficiently created, managed, and analyzed for detection of emergent issues. In addition, business intelligence can consistently be applied to enable optimal decision-making.
- *Use system engineering and research and development (R&D) personnel on all IPTs.* In companies like Boeing and Toyota, it has been documented that systems engineers and R&D personnel add value to the production of innovative and cost-effective products. While we are not building an airplane or a car, the

energy industry is continually dealing with extremely complex projects of massive scale that require even more coordination, integration, and testing of all hardware and software before deployment.

BETTER DATA MANAGEMENT

Energy companies experience difficulty managing the existing data that they have collected. In fact, so much information exists already that an increase in data from, say, automated meter intelligence for electric utilities or SCADA from inside underground oil and gas reservoirs raises the threat of "death by information." Even worse, as we have pointed out, the poor quality of much of the data being collected today poses a liability. CALM implementations (initially) spend a significant amount of time cleaning bad data (more than half-time). The following data management practices are helpful to enable the implementation of the CALM principles:

- Use open standards, open-source software, and COTS software (e.g., OpenGIS, Linux, Sun Java) whenever possible, to benefit from speedy advances from throughout the world. These provide ease of connectivity, collaboration, and communication.
- Use BPM software instead of mapping tools, because BPM has underlying connectivity to enforce process change.
- Add rules engines to capture, control, and change best practices, policies, processes, and procedures.
- Integrate planning and scheduling software into operations in order to flexibly repair or modify assets.
- Make everyone coordinate with the IMP and the IMS, which, when combined project to project, become the business plan of the company.

- Develop decision support software, such as machine and reinforcement learning algorithms, to better ascertain the health of assets and continuously improve the business.
- Foster scenario analysis and simulations that allow decisions based on optimal use of both qualitative and quantitative knowledge.
- Require a disciplined approach to decision-making by using business economics to justify and prioritize projects (e.g., if no-value, then no-go).

STRUCTURAL GUIDE FOR THOUGHT LEADERS

Significant changes are needed to fully implement CALM throughout the company. The process is usually begun through attempts to promote better collaboration between operational and engineering groups and the IT department—if, as in most traditional companies, they have not yet been well integrated into operations. IT departments, in particular, have for decades been cast off as a separate group of experts. In many cases, IT groups have been beaten down as a result of IT outages impacting operations and have thus become highly risk averse to innovation. As a result, IT has rarely been tightly coupled with operations and business decisions. They are very often brought into discussions too late in the game to collaborate effectively on new innovative designs and engineering.

It has been our experience that if IT personnel are fully integrated into operations, they deal more successfully with costs while producing better benefits, jointly. Otherwise, IT often takes second fiddle as the internal "vendor" that has to compete for work with outside suppliers.

Alternatively, if IT creates its own plans for managing the business, these often have little regard for or lack a full understanding of either the operational or engineering complexities required for success.

Fostering higher levels of collaboration between the organizations in areas like asset management and systems engineering will be critical to the success of CALM. As shown in figure 1–5, parts of the engineering and IT departments can collaborate within IPTs and thus be driven by the needs of the integrated system and its business operations. However, this requires significant teaching of business analysis skills to engineers and programmers alike.

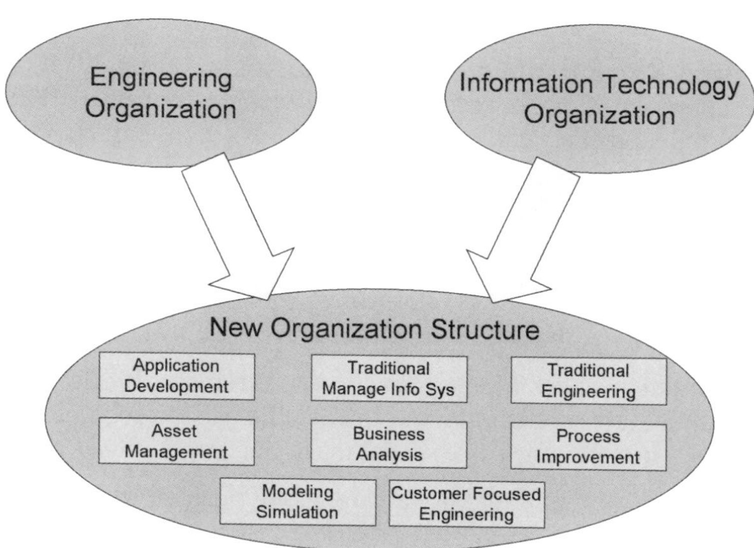

Fig. 1–5. Collaboration of engineering and IT departments with operations to enable better implementation of CALM.

For example, consider decisions for how best to deploy assets within and between projects. An integrated engineering/IT team can base its decisions on quantitative business intelligence by using an ISM to determine, through simulation, which of several strategies will be the best. The employment of simulations using integrated, accurate, and flexible models for the testing of scenarios and the computation of real options for design and decision alternatives will be how energy companies will be able to wean themselves from their reliance on current overbuilding of hard assets. Intelligent software models combined with real options encourage flexibility. The building of better business capabilities allows for better designs and deployment of assets. That, in turn, enables the following:

- Employment of dynamic redundancy in systems, to achieve enhanced reliability at lower costs.
- Modularity to replace one-offs and maintain real options for new designs.
- Expanded architectures to include sense-and-respond functionality—for example, intelligent SCADA controllers that can open and close valves and flip switches autonomously.
- Quick incorporation of future technologies, such as alternative energy resources that can augment the existing design basis of the deployed systems.
- Measurement of effectiveness through the use of mobile tools for reliability enhancement and rapid recovery.
- Movement away from deterministic measurements for the design of system reliability and toward the use of probabilistic analysis, to better measure and manage business risk.
- Operations and maintenance awareness of business risk.
- Enhanced knowledge of asset health and more effective maintenance on these assets.
- Understanding of uncertain events and extreme conditions, so that preventive actions can be planned and put in place before big, bad events happen.

With an ISM of the business in place, many emergent issues will become more manageable. For example, the ISM reduces employee errors and incidents of lack of compliance by mandating the use of software to monitor and control business processes in the field. Using the ISM to create simulations of business operations—to train employees, to experiment with better ways to improve operations, and to eliminate errors (some potentially catastrophic)—will be the other clear win that emerges. Additional best practices that create value are as follows:

- Software created to drive buddy-system compliance in order to ensure safe working conditions and control quality of operations in the field.
- Facilitation of continuous, mobile e-learning for procedural compliance and training.
- Design of software tools and mobile computers that prevent mistakes from being made by field workers. This will reduce system failures, some of which could be catastrophic.
- Prediction of the health of all key operating components and their criticality to continued service.
- Accelerated life-cycle testing of equipment with manufacturers, to enable better life-of-the-asset management.
- Appropriate consideration of environmental impacts and benefits.
- Optimized asset management using business intelligence software, to set appropriate priorities and plan preventive maintenance programs.
- Reduced fire, flood, storm, and other losses of physical assets through the modeling of physical constraints and the repositioning of highest-priority assets.
- Reduced potential for fraud, theft, or other criminal activity through the deployment of smart tags and intelligent algorithms to detect this activity and provide alerts.
- Suggestions of change by frontline workers. These ideas should be quickly tested using simulations and controlled experiments, to determine their effectiveness in driving out inefficiencies.
- Celebration of the success of effective changes in how work will be planned and performed.

TACTICAL GUIDE

With the implementation of CALM, a company's tactical strategies need to be augmented as well. A flat organizational structure with minimal vertical hierarchy will be important to create incentives that drive the teamwork necessary to accomplish the challenging goals. Providing personal incentives for improvement in business performance is dangerous, because striving to make the most money could come at the expense of the success of the team; the individual drive to make bonus money becomes a threat to team success and results in silos of information and knowledge that are counter to the objectives of CALM. Therefore, people in authority must foster the sharing of ideas by giving nearly identical compensation incentives that are based first on the company's performance, then on the team's performance, and finally on individual willingness to work as a team. The following are tactical additions and changes of importance to consider as part of a CALM strategy:

- R&D changes:
  - *Invest in innovative R&D to develop new products and services that enhance internal operational change.* R&D needs to be focused more intently on innovations in technologies and products that are catalysts for change.
  - *Create vehicles to commercialize new innovations, and use these to spread low-risk solutions to operations, maintenance, and engineering problems.* Technology development can come from contractors, universities with experience in the application of innovations, or consultants experienced in transferring technologies successful in other industries.
- Operational changes:
  - *Proactively develop and implement operational technologies that drive out inefficiencies and reduce risk.* Energy operations have

to deal with potentially lethal consequences of their actions, but that is no excuse for avoiding change that can mitigate these potentially catastrophic events in the first place.

- *Collaborate with partners to advance the development and application of innovative business intelligence software tools and services.* Creating an operationally efficient business requires significant commitment of brainpower, and that can be leveraged through strategic partners who are not true competitors.
- *Collaborate with partners to advance the use of smart algorithms.* Then, see that they are distributed throughout the system and used to enhance systems' operability and service to customers.

- Human resource changes:
  - *Train your existing subject-matter experts in the use of CALM techniques and tools.* They can then carry the torch of continuous improvement to their areas of influence.
  - *Hire experts in emerging technologies and business intelligence.* Stay abreast of external IT system and technical innovations that show the potential to create value in your business operations.
- Financial changes:
  - *Drive budget decisions for maintenance, operations, and new capital deployment that are based on business optimization models, not on last year's budget.* Employ business intelligence algorithms and use ISM simulations to rigorously evaluate real options for optimal expenditures and investments in your business.
  - *Make investment decisions that include realistic business risk analyses.* Business risk associated with uncertainties in the construction and operation of high-cost infrastructure is often given little attention and can lead to significant losses to enterprise value from catastrophic events that could have been simulated. Is your company playing Russian roulette and not aware of it?

  - *Modestly fund new opportunities that have been identified as providing significant real option value in the future.* Use scenario analysis to explore real options in the face of realistic appraisals of uncertainties in the world.
  - *Develop and implement investment policies and processes that exploit technology options in your business operations.* Recognize the requirements for continuous improvement of existing operations via the exploitation of new technologies. Otherwise, little progress will be made when resources fall short in areas where past solutions are no longer working.
- Regulatory changes (primarily for utilities):
  - *Ensure fair treatment for the application of technologies known to have considerable risk of failure.* It is imperative to communicate to regulators and governments the potential rewards to all stakeholders from the introduction of successful new technologies. Foster their understanding and agreement to invest in expenditures of this nature.
  - *Advocate clear regulatory policy for the transfer of successful new technologies to commercial partners and affiliates.* The need for maintaining successful new technologies that have been developed specifically for the company requires intellectual property protection and technology transfer for continuous commercial support of the inventions. With the advancement of technologies come opportunities to grow the company in these non–asset-intensive areas. However, this requires an assured mechanism for creation of shareholder value in the future. Creating an affiliate or subsidiary to further develop these technologies may require modest but high-risk investments.

DECISION-MAKING

Decisions about the deployment and maintenance of revenue-generating assets require that service to customers be paramount. If any business is to improve its value, critical factors relating to customer needs must be fully understood before actions are taken. In the future, the enterprise-wide ISM will be developed within the CALM implementation to simulate scenarios and calculate real options that do just that, thereby steering decisions toward optimal value creation. This value will be measured not only by short-term profitability but also by metrics for long-term risk avoidance.

To consistently take positive actions, decision-makers must first have situational awareness of the existing state of the business, including not only this customer impact but also the competitive and regulatory landscapes.

In nonlean businesses, including most of the energy sector, decision-making is not information driven, but is primarily handled by human intuition and experience. As proof, consider that energy businesses have found it hard to break out of cyclical profit-and-loss trends that are driven by well-understood uncertainties in supply and demand, as well as by impediments of regulatory and/or political change. Reliance on biased financial assumptions has led to billions of dollars of lost enterprise value—not from a lack of sophistication, but from inaccurate and false understanding about the uncertainties and risks facing the business. Energy companies must be brutally empirical in evaluating the true competitive environment. That is, such analyses often lack quantitative understanding of the true competitiveness of the firm's products and business capabilities.

Investment choices in infrastructure require continuous interpretation of constantly changing markets, which has been difficult to quantify because of the complexity of the industry. Thus, investment decisions are often based solely on the past experience of the decision-makers. That is, the requirements for continuous, 24/7 supply of energy drive most companies to focus on assets that they are historically comfortable managing. For example, to provide reliable supplies, an electric utility will likely emphasize increasing ownership in generation and transmission, rather than in better distribution to the customers.

Unfortunately, the future rarely follows the past in the energy sector. The only certainty is that things will change. It is surprising that uncertainties in the drivers of enterprise value in the energy industry have not been quantitatively evaluated using the algorithmic tools common on Wall Street. Instead, qualitative data are often used to arrive at quantitative decisions about the future. The qualitative approach has worked with on-and-off success in the energy industry for the past 150 years, but the economic signposts and geopolitical and environmental challenges energy companies face today make it clear that current strategies are suboptimal and can lead to missed opportunities—or worse. While the experienced eye plays a significant role in interpreting data, it must be tempered with the accuracy, flexibility, and reliability of computer analysis, as we will show in this book.

GOALS

The creation of an ISM removes the barriers posed by the silos, or stovepipes, inherent in the departmentalization of most companies. Integration enables lean uses of information for the creation of actionable knowledge. CALM strives to create such a lean management approach to running the company through the rigors of software enforcement. Each of the goals of implementation is interdependent with the others, and all are dependent on creating a collaborative operating environment (fig. 1–6).

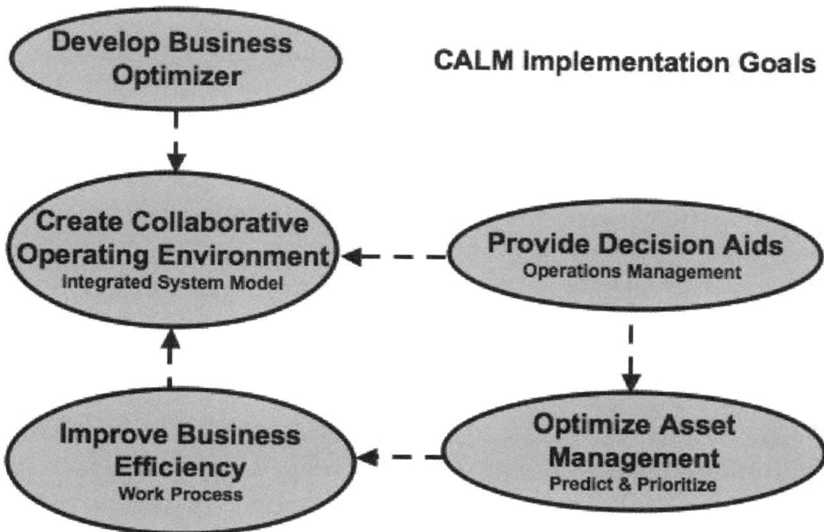

Fig. 1–6. Goals of CALM. Implementation is designed to create business capabilities that will enable the company to optimize asset management, business efficiency, and decision-making, all using a transparent, collaborative computer system for business optimization.

Creating a collaborative operating environment

A collaborative operating environment serves as the foundation on which a company will be able to transform the ways it conducts its business. Everyone in the company is then informed and working toward the same goals. Making quality information and analyses transparent and available to everyone enables the company to efficiently leverage past expertise against the changing dynamics caused by the uncertainty in the energy business. Creation of an ISM of the infrastructure of the entire enterprise, including every critical asset from source to customer, allows everyone to use the same information and understand the value of the same assets. An ISM that efficiently integrates all of this information greatly assists in decisions to purchase, design, configure, install, maintain, and monitor assets. This lean management approach is of paramount importance, since any manipulation of, reentry of, or discrepancy in information is, at the least, a pure waste.

Improving business efficiency

The real challenge will be to have your organization want to continuously improve the ways it works through refinements of business processes and practices. CALM uses a combination of BPM and chained matrices to capture the as-is state of the company. Flaws can be identified, and an optimal design can then be created to transform the company into a new, more efficient to-be state. Software is then brought in to improve not only efficiency but also quality control. During the conversion, accurate data are captured to correctly measure the state of the system, so that decisions that will improve value to the enterprise are consistently made. Improving business efficiency will be dependent on creating a collaborative operating environment.

Providing decision aids

Decision aids enable people to comprehend what is happening or will potentially happen to the system that they are trying to manage. Decision aids provide operators with the clarity to make sound decisions. In addition, it will be imperative that these decision aids provide a means to create effective actions based on these decisions in the future. Without successful actions, all the information, knowledge, and decisions fail to create value for the enterprise. Providing decision aids will be dependent on creating a collaborative operating environment and on optimizing asset management.

Optimizing asset management

Asset management reflects the desire to understand and master the risks of nonperformance of every component in the infrastructure. It will be used to measure enterprise value and manage plans to mitigate risks effectively. Many ways of measuring these risks are available. For example, we have developed ML algorithms to find a risk-adjusted susceptibility of impending failure for the electric distribution grid in New York City. Such a quantitative approach leads to opportunities to replace components that are most likely to fail, before they fail.

Contrast this with the business-as-usual, shotgun approach of the maintenance activities of most energy companies. They often replace components based on years of service or based on generic replacement programs, regardless whether the specific component is still performing well. Good predictions of which components need replacement provide the opportunity to mitigate the risk of catastrophic events and reduce costly non–value-added maintenance expenditures. If predictions and prioritizations are developed for the energy business as they have been for aerospace, medical, and consumer industries, then we have the potential to also multiply the effectiveness of our capital spent on preventive maintenance and asset management. Optimized asset management relies on improvement of business efficiency.

Developing a business optimizer

Business optimization will be at the heart of improving enterprise value in the future. It requires informed analysis of resource deployment, real options, risks, and opportunities in the future. A business optimizer is an integrated, company-specific set of algorithms that enables the best decision-making to meet the corporate strategy. How do you create the budget for the work your company will do next year? Is your strategy achievable given how you are planning to deploy your resources? How do you know that your use of these resources is going to be effective? Developing a business optimizer will be dependent on creating a collaborative operating environment that jointly can answer these questions in advance of the investments.

CALM uses state-of-the-art statistics and computational analysis techniques to more accurately measure system risk. This better quantifies the reliability of service to customers, as well as the business risk associated with this service. A more efficient redesign of the "pipeline of services" from sources to customers for electricity, gas, oil, and renewable energy resources comes from the implementation and use of the ISM. The implementation of BPM, ML, and better asset management feed the business optimization engine within the ISM. These components are combined into a collaborative operating environment that will be distributed to all workers, from the lowest field worker to the highest executive. Reduced business risk to the company and better service to the customer immediately follow.

IMPLEMENTATION

It will be challenging to implement CALM in field-intensive industries because statistical tools and measures do not all exist to efficiently model and track progress. For example, current IT standards in the "field asset" parts of the energy industry, such as upstream exploration and production of oil and gas or distribution of electricity, do not routinely collect measurements of component reliability. *Loss of load probabilities* and *loss of load expectations* are examples of failure models used in bulk power delivery that aim to measure risks of outages. Unfortunately, most of these failure models do not extend to the entire electric system because they are unable to deal with complexity. In contrast, these statistics are tracked routinely in the "factory" parts of the industry, such as refineries and petrochemical and power plants.

In addition, accurate failure models for these field asset components have not been developed for continuous monitoring, reliability analysis, and preventive maintenance. A good measure of whether such models exist is if control-centers are used for process management and maintenance of the field assets. In the energy industry, electricity generation, transmission, and distribution and downstream refining and processing use control-centers much more extensively than upstream oil and gas production. Where missing, there are significant opportunities to improve enterprise value from the introduction of the ISM to connect control-centers more directly to business decisions (fig. 1–7).

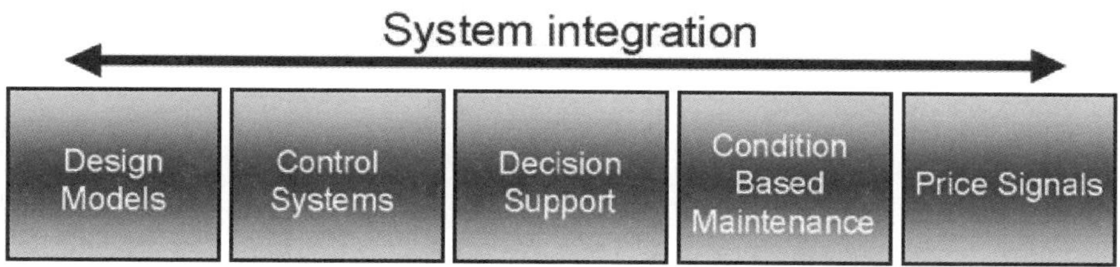

Fig. 1–7. Integration of systems. Better design, control, operations, maintenance, and economic decision-making result when everyone in the company uses the same model.

The oil and gas business is less regulated but more global and complex than the electricity business (both industries are everywhere, but the coverage of any electricity company is more local). The irony is that the least profitable portions of the overall oil and gas business, downstream refining and marketing, are the most efficiently run because the thin profit margins require process rigor. Conversely, the most profitable field components of that business, exploration and production, are the least efficiently run and are most susceptible to wild swings in performance.

Preventing train wrecks alone would justify the change to CALM in all capital-intensive portions of the energy business. Consider the production of ultra-deepwater oil and gas: Two multibillion-dollar offshore platforms have been severely damaged in the past few years (fig. 1–8). Almost everything so far has been overbuilt, including both of the mammoth projects shown in figure 1–8, and yet, this wasteful and inefficient overbuilding has done little to insure survivability, let alone sustainability.

Fig. 1–8. Two capital-intensive field assets that would have benefited from CALM processes and technologies. The ultra-deepwater production facility (top) was the largest floating oil platform in the world prior to its sinking, after an explosion, in March 2001 (AP photograph), and the largest moored semisubmersible in the world (bottom) was found severely listing following a hurricane, in 2005 (U.S. Coast Guard photograph, by PA3 Robert M. Reed).

Only through one ISM of all assets and customers, combined with advancements in computational learning algorithms that offer the hope of truly optimizing business decisions, can a company efficiently find the information needed to improve decision-making in order to prevent such big, bad events in the future. The complex mixture of real and simulated data and model integration needed to create an ISM in the electricity business is represented in figure 1–9. These models currently require manual entry of data from system operations. Contingency analyses that measure the reliability of delivery to customers now occurs only after failures. How do we minimize risks from future outages should there be many more transmission lines, or more substations, or more generation, or more distributed generation, or more of all of these?

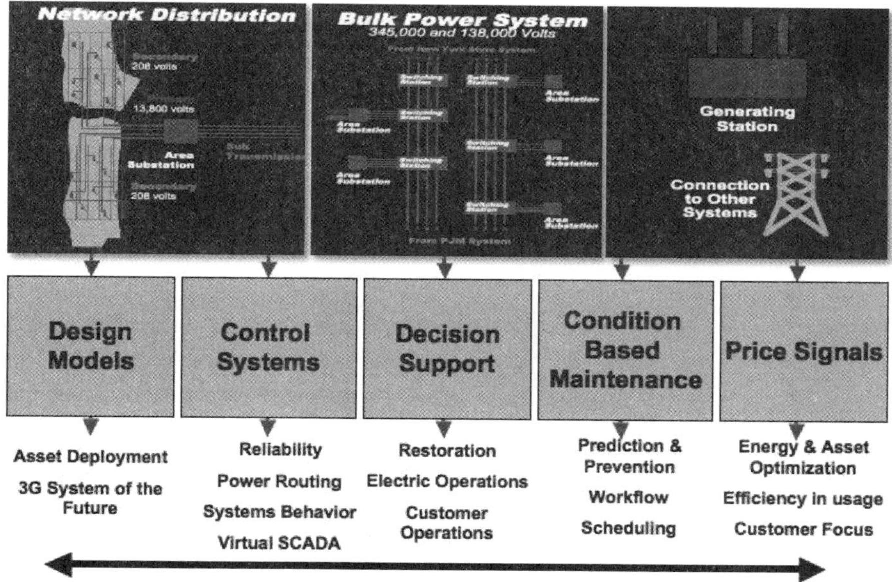

Fig. 1–9. Process integration by the ISM, requiring the crossing of three traditional business silos in the electricity industry—power generation, transmission, and distribution.

The ISM will simultaneously compute asset deployment strategies, while ensuring system reliability, and best maintenance practices. At the same time, energy throughput will be optimized, and prices are tracked. The continuous monitoring of system performance and simulations of next-worst events that might be imminent allow corrective actions to be taken before train wrecks happen.

Management's capability to meet or exceed year-over-year expectations can be enhanced through better measurement, analysis, and simulation. CALM processes improve value delivery to customers, reduce costs, utilize capital better, avoid overbuilding, maintain focus on value generation, and reduce business risks through the use of modern business technologies. Consider the following risks and uncertainties of the energy business:

- Regulation:
  - Governmental actions exerting a negative influence on earnings
  - Failure to comply with regulations
- Market:
  - Assets becoming less valuable because of competition from other sources of energy
  - Prices of oil, gas, and electricity drastically changing
- Operation:
  - Failure to provide value to the customer
  - Failure to prevent harm to the public
  - Failure to identify and prevent major incidents
- Reputation:
  - Negative opinion of the public toward the company
  - Negative opinion of governments toward the company

Computational sciences can be used through an ISM of the business to measure, monitor and control these risks today. We will provide examples of what we have done to date and point to potential solutions that make managing these uncertainties an opportunity to grow.

By the end of this book, we will have examined how the CALM methodology is going to enable future electricity smart grids with the efficiencies necessary to serve urban expansion in the 21st century. The same can be said for the oil and gas industry as it deals with both dwindling geological supplies and emerging renewable resource competitors. This industry will need proven computer intelligence software to respond to the new exogenous drivers in its business.

Developing the business capabilities of CALM will dramatically improve the business operations of energy companies. The opportunity exists to design and then operate all energy businesses in a way that will allow companies to reduce the amount of contingency hardware required while enhancing customer satisfaction, reducing risks from unreliability, reducing society's carbon footprint, and improving shareholder value. CALM makes these multiple win-win-win objectives achievable, even though they are often considered to be in conflict.

NOTES

[1] E.g., see Department of Energy. http://management.energy.gov/earned_value-value_management.htm.

[2] Giles, J. 2005. Internet encyclopaedias go head to head. *Nature.* 438: 900–901.

2 HISTORY

CALM is software-controlled, lean systems integration that drives innovation toward breakthrough cost and cycle-time savings and produces better use of capital. Lean management is already used throughout the communications, chemical processing, aerospace, and automotive industries, as well as other high-technology manufacturing and processing industries. However, it is not yet widely used in the energy industry.

This book establishes the background, logic, processes, and systems changes required for a company to migrate to the next generation of lean management. That next level entails computer-aided enforcement of actions and results, to transition from goals to actual innovation and continuous improvement in any business environment, no matter how uncertain.

To understand CALM, we need to first understand how it has grown on the "shoulders of giants" in that lean processes and techniques are already dominant management technologies with a wide diversity of applications and successes. Extensive Internet resources on lean management are accessible through search engines. For example, Google will return more than a million results for a "lean management" search. Wikipedia is another excellent place to begin to understand the many variations of lean concepts. More than 2,000 different wiki articles refer to lean principles.

We begin with how the many facets of lean management fit together to form the CALM principles discussed in this book. Figure 2–1 is not meant to represent a synthesis of the lean world now, but instead we show how the overlapping goals of the many different lean perspectives fit together in the many companies and industries that actively use lean methodologies today. We term the combination of quality and work practice management *lean and mean,* because if lean transformations are administered solely based on these two principles, performance incentives (bonuses [the lean]) and disincentives (firings [the mean]) drive workforce motivations. If work practices and value management are mostly used, *mass,* or focus on physical asset management, dominates in the form of capital, operations, and maintenance cost cutting. If quality and value management alone are the dominant focuses, the methodology known as *Six Sigma* has emerged. A search of Google will turn up almost 10 million Web sites about Six Sigma. Only the combination of quality, work practices, and value management into a single integrated, computer-aided, win-win-win environment for employees, shareholders, and customers will produce a lean business that is flexible enough to deal with the uncertainty inherent in field-intensive industries.

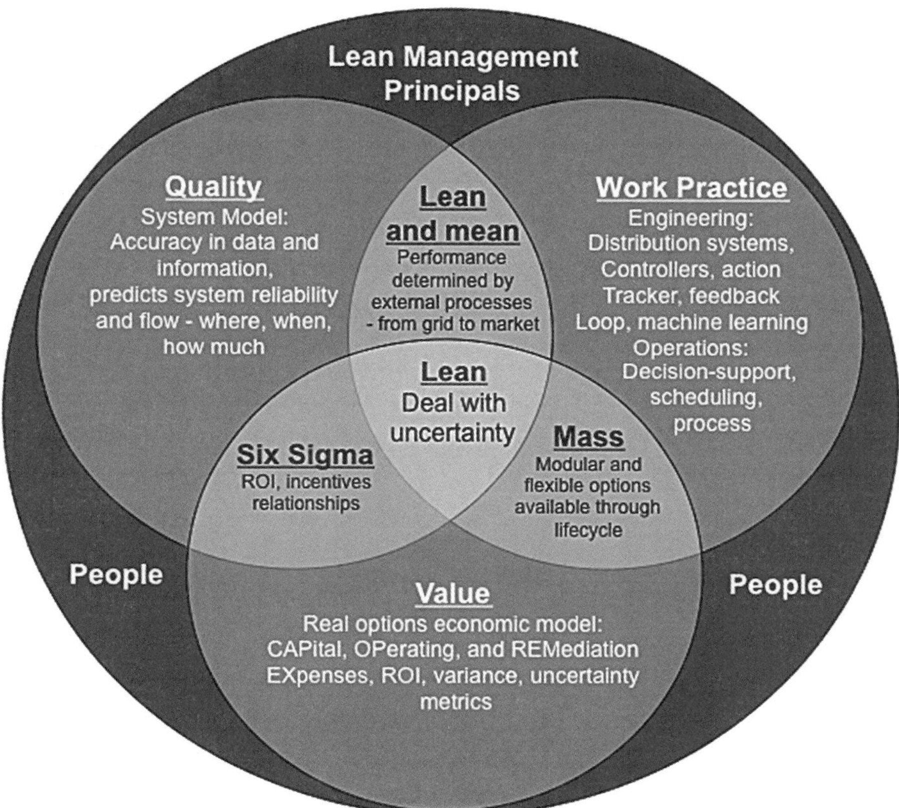

Fig. 2–1. Lean principles. CALM implementations concentrate on simultaneous improvement in three overlapping spheres of influence: quality, work practices, and value management.

In addition to quality, work practices, and value management, there is a fourth major component of lean management that is all encompassing, and that is the people involved in the work. These four components of lean management overlap in three subsystems—function, information, and data flows—that interact to stimulate processes, produce new designs, create models for knowledge sharing, devise systems and architectures, and organize networks of influence within a lean company.

General Electric (GE) is famous for the concepts of *Six Sigma* and *lean and mean,* Boeing made their mark managing *mass* (using modular and flexible assembly options), and Toyota put it all together into a truly lean business culminating in the success of the Lexus automobile. Boeing in turn took the Toyota model and added computer-aided enforcement of lean methodologies throughout the manufacturing process. In this book, we describe these and additional futuristic techniques that will allow lean management to migrate from the factory floor to operations that are *field oriented*—that is, affected by additional uncertainties, such as weather, climate, interdependent equipment failures, and threats from exogenous forces (e.g., enemies shooting at you) that are not present in indoor factory assembly lines.

INTEGRATED DEFINITION

Lean management is a methodology for efficient enforcement of process rigor and discipline to dramatically cut costs and improve cycle times of all operations of an enterprise. One of the major sources for its outgrowth was *integrated definition* (IDEF) modeling in aerospace manufacturing, pioneered by the U.S. Air Force in the 1970s. IDEF is a methodology designed to model the end-to-end decisions, actions, and activities of an organization or system so that costs, performance, and cycle times can be optimized. IDEF methods have been adapted for wider use in automotive, aerospace, pharmaceuticals, and even software development industries. Google returns more than 200,000 Web sites that refer to IDEF, including many .mil sites.

IDEF methods serve as a good starting point to understand lean management processes. The IDEF process always begins by mapping the as-is functions of an enterprise, creating a graphical model, or road map, that shows what controls each important function, who performs it, what resources are required for carrying it out, what it produces, how much it costs, and what relationships it has to other functions of the organization. IDEF simulations of the to-be enterprise have been found to be efficient at streamlining and modernizing both companies and governmental agencies. An excellent IDEF methodology Web site is maintained by the U.S. National Bureau of Standards, through a subcontract with Knowledge-Based Systems, Inc. (see http://www.idef.com/).

GE

Motorola developed the Six Sigma methodology in the mid-1980s as a result of recognizing that products with high first-pass yield rarely failed in use. At GE, it has grown into more general lean principles that are rigorously enforced throughout the organization. Software is used to make the entire manufacturing system transparent and measurable, whether it's a lightbulb, an electric generator, a jet engine factory, or a power plant. GE requires process mapping of the as-is condition of whatever system is to be improved, establishment of baseline metrics, identification of where the waste is occurring, and planning of the improved to-be process—all shared interactively on the computer before change is authorized. Then, management controls the implementation of the innovation plan, with constant reviews of performance metrics along the way.

As good as its technologies are, GE does not distinguish itself through innovation so much as through execution of the systems integration processes necessary to manage innovation. Whether it is a new product or the manufacture of reliable light bulbs or jet engines, they are very good at producing quality products. Through continuously scoring the performance of their people and rewarding the top performers and firing the bottom 10%, they have become famous for creating a company that is both highly successful and lean and mean.

TOYOTA

The most famous lean root is the Toyota Process Management model, with their Lexus division's "relentless pursuit of perfection" being its most prominent slogan. The Japanese derivatives—*Kaizen* (continuous step-by-step improvement through teamwork), *Kaikaku* (rapid improvement through shared efforts), and *Jidoka* (balanced responses to external forces)—can be found at the base of the process improvement methodologies of most major innovation efforts today, both inside and outside the automotive industry. Toyota begins with the mapping of existing processes, so that a plan for migration from unhealthy to healthy processes can be planned (see top of fig. 2–2). Toyota's methodology then is to turn attention to *bottom-up* process management, consisting of *learning steps* in which improvement skills and knowledge are rigorously taught to all employees, followed by *standards building,* so that metrics of improvement can be mapped. Only then does the "do" action start (e.g., this is where we perceive Six Sigma fits with Toyota Process Management).

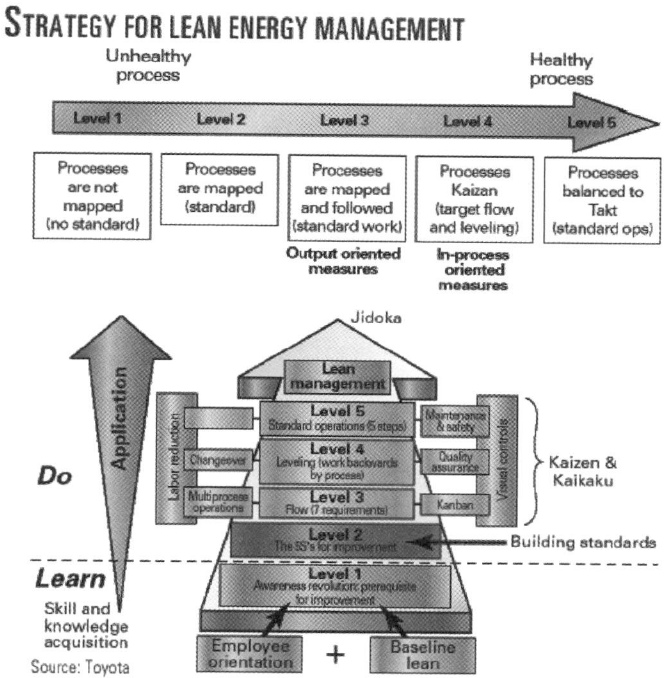

Fig. 2–2. The CALM strategy for lean energy management (shown horizontally). The origins of this strategy can be traced to the Toyota Way, (shown vertically). ***Takt*** is the maximum time allowed to produce a product in order to meet demand.

In particular, Toyota made significant improvements in dealing with its subcontractors through lean systems engineering. Toyota realized that everything—from just-in-time inventory delivery to total quality management and rapid adoption of new innovations—depended as much on the performance of its outside suppliers as on performance within the company itself.

At Toyota, corporate investments in new technologies include not only acquisition and venture investments but also loans and, sometimes, outright gifts to suppliers to get them to buy into the Toyota lean automotive system. Why go to the extra expense? It has allowed Toyota to instigate metrics and performance standards that can be tracked all day, every day, throughout the greater organizational system that includes their subcontractors.

Boeing, Lockheed Martin, and countless other great corporations throughout the world have adopted and adapted Toyota's model. In fact, Toyota invited General Motors, Ford, and Chrysler to come to Toyota City to learn lean automotive systems, believing that a better overall industry would ultimately benefit Toyota as well. Even though they came by the thousands, U.S. automobile manufacturers have still not quite "got it," as can be seen by the quality and sales of their products. However, the hiring of a top Boeing executive to lead Ford is a sure sign of lean improvements to come.

A major innovation missed by the Americans is that the rigor increases steadily from level 1, where processes are mapped, to level 5, where processes are scripted and controlled often to the minutes it takes to do each car assembly task. Thus, continuous improvement is presented within Toyota as a ladder (fig. 2–2). Toyota describes how to manage the learning of lean processes by use of its Jidoka strategy, defining these five levels of growth (fig. 2–2), but enforcement is a much harder problem.

The levels of growth in Toyota's model are as follows:

- *Level 1* requires the benchmarking of the existing manufacturing processes, whatever they are, in order to create a baseline to measure future progress. In addition, employees and subcontractors are all introduced to lean process theory. They are challenged to stop measuring specific actions and instead think of each and every process in terms of the whole system that they are producing. It is only the cumulative effect of all those actions that results in a quality product.
- *Level 2* requires that software tools be put into place, to build and enforce standards and identify and eliminate waste in material, machines, effort, and methods.

- *Level 3* tools escalate to the introduction of a common, three-dimensional solid model for all to use to standardize and streamline work.
- *Level 4* introduces a continuous improvement plan.
- *Level 5* finally achieves lean management.

BOEING

Perhaps the best-developed evolution of the IDEF model beyond Toyota is at Boeing. Their project life-cycle process has grown into a rigorous software system that links people, tasks, tools, materials, and the environmental impact of any newly planned project, before any building is allowed to begin. Routinely, more than half of the time for any given project is spent building the precedence diagrams, or three-dimensional process maps, integrating with outside suppliers, and designing the implementation plan—all on the computer. Once real activity is initiated, an *action tracker* is used to monitor inputs and outputs versus the schedule and delivery metrics in real time throughout the organization. When the execution of a new airplane design begins, it is so well organized that it consistently cuts both costs and build time in half for each successive generation of airframe. And, of course, it is paperless. Boeing has found that these cost and cycle-time savings can even be achieved on one-of-a-kind production projects, such as the X-32, X-45, and X-50 experimental air vehicles.

Lean management progress came with more than 20 years of growth through high-impact implementation opportunities, such as the design and construction of the Boeing 777 (fig. 2–3). Such high-risk applications required a commitment from top executives that allowed more than a billion dollars of expenses before the first successes were realized. The manufacture of the 777, touted as the paperless airplane, really represented a make-or-break evolution of the company. Boeing was to become lean or go out of business.

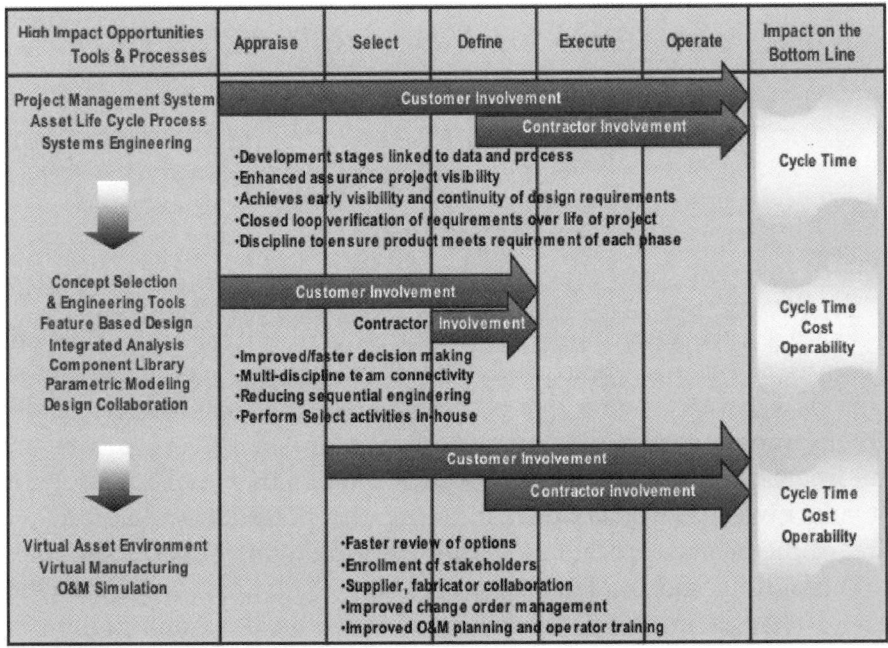

Fig. 2–3. The Boeing Way. They strive to integrate contractors and customers in all phases of the asset life cycle, from conceptual design, through manufacturing, to operations and maintenance.

Boeing created a complex lean management process called *define and control airplane configuration/manufacturing resource management* (DCAC/MRM). The process was built with the help of the operations research and computer sciences departments of the University of Pittsburgh. The manufacture of the 777 was ultimately a success, and it became the precursor to succeeding generations of CALM at Boeing. Boeing is four generations beyond that airplane now, and they have succeeded in cutting the time *and* cost by 50% for *each* new generation of airplane by using CALM. Boeing's successes in conversion from inefficient silos of manufacturing to a lean and efficient operation have become legendary (fig. 2–4).

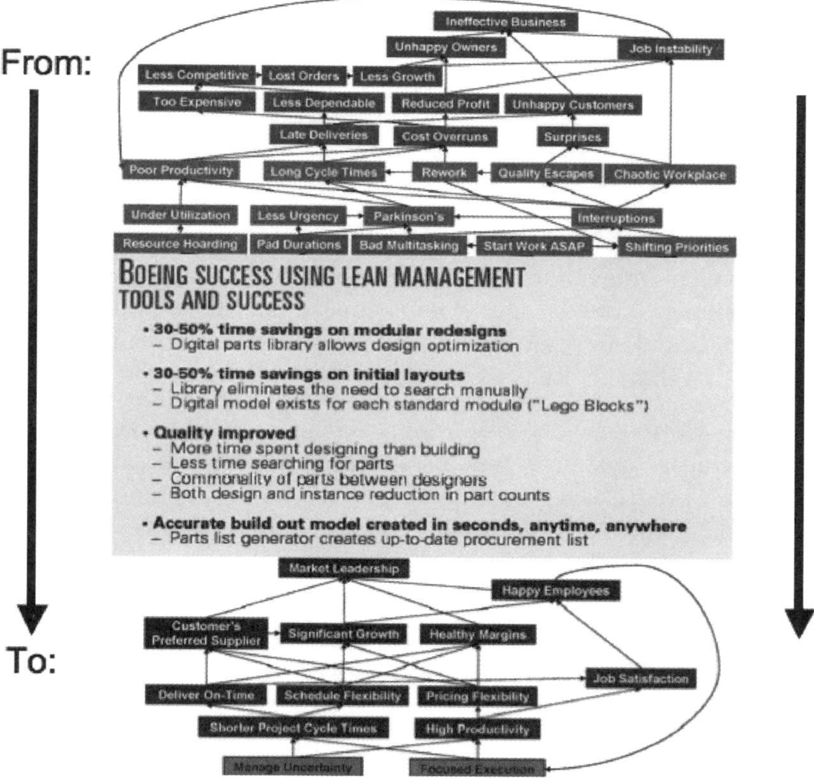

Fig. 2–4. Benefits of moving from inefficiencies to efficiencies. As a result, Boeing has experienced a decided improvement in business performance and has moved into a dominant position within the aerospace industry.

Phantom Works, Boeing's R&D equivalent to Lockheed Martin's Skunk Works, now has an entire division devoted to lean and efficient design of processes and tools. The following quotation is from Boeing's

DCAC/MRM entry into the *Computerworld* Smithsonian Awards in 1995 (the awards honor the creativity and inspiration of those who use modern technology to benefit people; this quotation is now part of the permanent archive at the Smithsonian Institution's National Museum of American History):[1]

> Since the 1940s, when the Boeing Company pioneered the mass production of large airplanes, we have built on our tradition of engineering and technical excellence. Over the decades, this tradition established us as the world's leading manufacturer of large jet airplanes. Our engineers used the latest technologies to develop innovative airplane components and systems driven by customers' needs. Our competitive dominance continued, but it was based more on technologies applied to airplane component design than on efficient processes and systems.
>
> However, by the late 1980s, we realized that the business systems we used to manufacture airplanes had not progressed as quickly as our airplane technology had advanced. We made several attempts to improve our processes, but progress was slow, largely because the proposed improvements were local rather than companywide. It was clear that, to maintain our market position and keep a stable, well-motivated workforce, we needed to revolutionize the way we do business.
>
> We studied other companies around the world and realized that, to stay ahead of our competition, we had to plan for the future. Our processes had to be flexible enough to evolve as the market changes. At the same time, we needed to reduce production costs and defects by 25 percent and time-to-market by 50 percent while completely satisfying our customers. In 1994, the president of Boeing Commercial Airplane Group chartered a breakthrough production strategy that attacked several key problems simultaneously.

The resulting Define and Control Airplane Configuration/ Manufacturing Resource Management (DCAC/MRM) project simplifies the processes and systems that we use to configure and define the components for each airplane, to build and assemble those parts, and to manage the data. DCAC/MRM is the result of a dynamic cooperative effort between us and our information technology providers. DCAD/MRM integrates new suites of commercial off-the-shelf software applications and products into a highly flexible and powerful system that allows us to select the right process for each job. The processes and software of DCAS/ MRM help the diverse organization of our large enterprise to function like an integrated network of small businesses.

Before the implementation of DCAC/MRM, the very size of our company was a detriment to its efficiency. The Boeing Company had grown and divided into separate organizations, each with its own products, goals, and objectives. Business processes and interlocking information systems evolved to meet the needs of these organizations instead of the needs of a unified Boeing enterprise. We used more than 800 computer systems to design, manufacture, control, and support airplane components and assemblies. Most of these systems were aligned according to function, were not integrated, and did not communicate with each other.

To add to the complexity, some of these systems were up to 40 years old. Our complicated processes and systems required new design and planning every time we built an airplane, even though up to 75 percent of the airplane components are common to all airplanes of the same model. Furthermore, the list of components, or bill of material, created for each airplane had to be converted for or manually re-entered into as many as 14 bill-of-material systems.

Compounding the problem was a configuration control system that assigned a customer-specific identification number to each engineering part drawing. Each time we released a drawing, company engineers marked the drawing with the customers' identifier. Drawings were marked manually with the customer identifier even when the part had been built and used hundreds of times and even when we were not making any changes to the part. Drawings had to be re-marked when airplanes changed position in the production line to make sure customer configurations were accurately maintained. This over-processing caused bottlenecks in production, as employees did unproductive, repetitive tasks. Of the engineering time spent revising a drawing this way, we estimate that up to 95 percent added no value to the final product.

The future evolution of CALM will require that the tools and techniques be modified to account for the uncertainties of field operations outside pristine manufacturing facilities, such as those used by Toyota and Boeing.

FIELD INDUSTRIES

There are many similarities between experiences in the automobile, computer, and aerospace industries and those in the energy business. All require large-scale systems integration of complex engineering processes to produce profitable products. They all involve multiple suppliers that are global, and there are many common suppliers to many owners. Lean management provides for earlier engagement of IPTs, improved access to new technologies, earlier rigor applied to "go/no-go" decisions, and an enhanced resource base and skill level for managers and engineers. In this time of the graying of manufacturing industry personnel, a lean energy revolution also opens up entirely new employment possibilities: aerospace, computer, and automotive engineers and managers can transition to field industries like the energy businesses. These lean thinkers would bring a fresh outlook to the elimination of customization, complexity, and interface conflicts in the more staid and traditionally managed field industries, such as electric utilities and oil and gas production.

Lean management lowers costs through more commonality in supply, compression of cycle times, keeping all projects on time and on budget, control of risk, innovation matched to needs, and improvement of first-year operability of new equipment. Benefits documented by the U.S. Department of Energy (DOE) include continuous improvement in reliability, leveraged inventories to reduce risk, cuts in waste and spoilage (e.g., alpha uses of software codes produce one-year paybacks), use and reuse (from data to packaging), reduced errors in data entry, and engineering/IT integration. These improvements have been realized primarily by the lean aerospace and automotive industries, but the DOE believes that they can be adapted successfully into a lean energy management model (see table 2–1).

Table 2–1. The initial and then continuous improvement benefits that Boeing has documented from the adoption of their version of CALM for business and manufacturing processes.

Benefits Realized From Lean Management in Far Field Industries

		Intial conversion	**Continuous Improvement**
1	Labor productivity	Double	Double again
2	Production throughput times	90% reduction	50% further
3	Inventories	90% reduction	50% further
4	Errors in final deployment	50% reduction	50% further
5	Scrap material	50% reduction	50% further
6	Time to market	50% reduction	50% further
7	Engineering changes	50% reduction	50% further
8	Fabrication costs	50% reduction	50% further
9	nonperformance costs	70% reduction	50% further

Plus paperwork virtually eliminated

Challenges that lean management overcomes are dealing with the risks, opportunities, and uncertainties of the technological frontier while reducing the costs and cycle time necessary to exploit new innovations (table 2–1). Lean management processes and tools have created fundamental improvements elsewhere because the behavior of the entire system was fundamentally changed. Such is needed in the energy industry, although it will be more difficult because of the added complexity of weather, market uncertainty, and global reach.

Energy industry management will encounter a predictable set of reactions to attempts to impose lean concepts, processes, and tools from the aerospace and auto industries. The following human responses were repeatedly encountered and documented during lean implementations from company to company in those industries:[2]

1. *You don't understand our business: ours is harder; ours is different.* This usually comes from experts in the company who believe that what they do is really different from most other businesses in terms of technology and engineering. However, what they miss is that the capabilities needed to improve performance are consistently the same from company to company, whatever the business. Only slight variations are needed to fit each specific business or industry.

2. *We don't need to be lean; we just need to do what we do now less expensively.* As an analogy, consider a town-to-town journey by stagecoach in the Western frontier country of the 1800s that takes 8 hours. With lean, we think how to create value, which directs us to find ways to innovate by shortening the path, greasing the wheels, lightening the carriage load, or improving its design. The results of this thinking enable the horse to easily gallop for longer distances. This adds more value to the passengers on the coach, since they are less burdened by the travel time—a win-win proposition for everyone, even the horses. This is how Wells Fargo was born. The less expensive way would have been to increase the number of passengers on the carriage to increase revenue passenger miles. The passengers are squeezed together like sardines,

and the carriage needs more steel to hold the extra passengers; consequently the carriage is made more expensive, and the horses have to slow down because of the extra weight. The 8-hour journey stretches to 12 hours, and the horses eventually collapse in exhaustion from the overload and hours of travel. In addition, there were threats such as unpredictable weather and attacks from Indians. Which stagecoach would you rather be on? It's the difference between working smarter and working harder over a longer period. The smarter approach leads lean management thinking. In most companies, your dedicated employees are already working harder and longer. To get dramatic improvement, you have to become smarter in the ways you plan your work and work your plan.

3. *We are already using lean: we're doing* x *and* y; *only* z *is missing.* There is truth to these comments in that in all companies, there are pockets of innovation and efficiency improvement. The problem is that *x* and *y* are not company-wide, and without *z*, they rarely lead to lean benefits throughout the company. In most companies, these innovation leaders often leave in frustration for jobs in leaner companies. Individual attempts at lean innovations are usually unsupported, with no incentives, and they only give excuses to conservative management for why continuous, dramatic improvement is not possible.
4. *Only design and build teams will benefit from lean processes—not human resources, not finance, and certainly not accounting or customer operations.* The reality is that from a value perspective, many of these back-office processes have historically lacked the economies of scale in their respective silos to command investments in lean practices for efficiency improvement. This makes them more lucrative places to extract value from lean management.

Lean management will take time to implement fully in any company—and particularly in an industry like energy that does not have a history of lean development. The preceding reactions must be worked through to resolution, though, and examples of success by early adaptors will become critical teaching tools to overcome the risk involved in conversions at other companies. Awareness that train wrecks like those that have previously occurred can be prevented in the future and a realization that technology alone will not produce the continuous, step-change improvements promised by lean management are two attitude barriers to overcome. Fundamentally, lean management is a people process, and the human and institutional attitude changes should be a primary concern.

The deepwater oil and gas industry provides the long-term physical assets that can result in massive train wrecks with, for example, several recent examples of billion-dollar production platforms that have sunk, taken on heavy damage in a hurricane, or been dropped to the seafloor during an installation mishap. Three key CALM technologies can contribute to both improving acceptance by the workforce and increasing performance while preventing train wrecks. We order these from tools available today to those being developed and tested for future use throughout the enterprise:

1. Use computational machine learning (ML) to improve performance by scoring the results of field actions so that continuous improvement can be achieved and documented. Also, lean is data intensive and ML is used to convert data to information.

2. Utilize real options to ensure that decisions are valued and acted on as uncertainties play out in the future real world of field operations.

3. Reinforce positive outcomes and eliminate negative results in operations by using adaptive, anticipatory, stochastic control concepts to react to out-of-specification performance metrics and recommend optimal actions in real time.

When the manufacturing and deployment of field services and equipment is eventually managed using CALM methodology in the future, efficient, risk-mitigated, and cost-effective operations follow. Next, we turn our attention to the lean tools and techniques that are available now and those that are under testing and development for future deployment that hopefully will enable successful CALM transformation throughout the future energy industry.

NOTES

[1] *Computerworld* Smithsonian Award in Manufacturing in 1995: Boeing Commercial Airplane Group, 777 Division, Computing and the Boeing Design. Their case study is archived by the National Museum of American History in Washington, DC, a part of the Smithsonian Institution. See http://www.boeing.com/news/releases/1995/news.release.950614-a.html.

[2] Murman, E., T. Allen, K. Bozdogan, and J. Cutcher-Gershenfeld. 2002. *Lean Enterprise Value.* New York: Palgrave.

3 COMPONENTS

WHAT KEEPS CONTROL-CENTER OPERATORS UP AT NIGHT

During our cumulative careers, we have had extensive experience with operations in a wide variety of energy industry control-centers, as well as those in military, aerospace, chemical, and Internet industries. Over the years, we have asked operators and shift managers what keeps them up at night. From those in electric system power plants and transmission and distribution centers, to those in oil refineries, petrochemical, liquefied natural gas (LNG) plants, and offshore production facilities, the overriding worry that we encountered was that there are just too many surprises. The major management and workflow issues that were ranked as most important can be classified into three general themes:

- Information:
  - We don't know what's happening on our system in real time.
  - We measure most of it but don't have the data at our fingertips when we need it.
  - We lack the right information, when and where it is needed. There are too many missing data points, so we can't rely on the information that we do have.
- Understanding:
  - We don't completely understand how our system works.
  - Real-world events can't be predicted, nor can outages be prevented.
  - We don't know when, where, and why components fail.
  - There is not enough evidence of root cause to take preventive maintenance action—so we replace everything we can think of that might fail.
- Decisions:
  - We would like more feedback that we are making the right decisions at the right time.
  - One hand does not know what the other is doing.
  - It is hard to keep track of all that's going on.
  - We are not analyzing the data that we already have.

As with other industries or agencies that put workers and the public in harm's way to perform services necessary for the general good (e.g., the military and aviation), the decisions that energy control-center operators make have serious consequences, and we have found that, uniformly, the operators take their responsibilities extremely seriously. Loss of service, loss of reputation, and loss of life could result. A bad decision might necessitate too many reworks downstream as the problems cascade. This can waste too much in terms of both lost effort and time. Like the problem with medical patients so beautifully described by Shannon Brownlee in *Overtreated,*[1] energy control-center operators and managers are wasting resources overtreating problems that, if they were to happen, would be big, bad events, instead of fixing what is most susceptible to impending failure. For example, *overattending* to a failed landing-gear indicator light led to the Eastern Air Lines L-1011 crash in the Florida Everglades on December 29, 1972.[2]

It has been proved in other industries that mapping how operators deal with everyday events and how they relate to other people during crises is essential to good control-center decision-making. All modern control-centers have massive amounts of software that variously attempt to take the pressure off the operators and help them better manage risks while optimizing performance. The massive amount of training that control-center operators go through was really the first step in implementing CALM. Tapping the operator experience base and learning how they integrate information into business decisions comes next. Solutions that integrate information and push next-worst contingency estimates ranked by not just how bad things might get but also by what is most at risk and likely to fail next are one step away from being gathered into an ISM.

In order to reduce stress, CALM attempts to unite these legacy systems to enforce quality control and automate the push of only the data or patterns of data that are most in need of attention to control-center operators. The next steps are to add a PM to generate simulations of the business and exogenous drivers, combine it with BPM software and ML algorithms to better enable predictions of impending failures. The ISM holds the promise of combining these with legacy systems to create control actions to continuously mitigate risks and improve efficiency.

Transformation of the business will require many years of effort within any company to integrate the software components of CALM with the existing legacy software systems. Extension of CALM from control-centers to other operational and support divisions of a company then requires systems engineering in addition to traditional, engineering-focused project management because of the large, complex development and integration efforts required. We introduce CALM-based systems engineering processes in the next chapter. Eventually, we will see CALM operations throughout the energy industry making operators smarter, supported by the continuously running ISM, with its PM, BPM, and ML information processors introduced in chapter 1. In the successive sections following, we first identify computer technologies and techniques of each of the ISM, PM, BPM, and ML that are operational today and then clearly identify those that are still in development as parts of the grand vision of where we forecast that CALM implementations are headed.

INTEGRATED SYSTEM MODEL

Modern advances in business intelligence, computer systems, process management, and modeling have made significant dents in driving inefficiencies out of businesses already. Most of these solutions were created in the past 10–15 years, however, and many companies in the energy industry are famous for their traditional ways and late adoption of new business technologies.

Many famous corporations have based their entire business processes on utilizing these computer advancements, and they have disrupted their competitors in virtually every case through the creation of ISMs. Some well-known companies that make extensive use of ISMs—in addition to Toyota, GE, and Boeing (the founders of lean management discussed in chap. 2)—are Apple, HP, and Southwest Airlines. In particular, the creation of an ISM is common to success in transportation infrastructure businesses.

Our vision for the future of the energy industry is that they will build ISMs that orchestrate the work performed on their systems, whatever it is, and continually run various simulations using PM, BPM, and ML inputs to understand the constraints acting on the system (fig. 3–1). Employees will comprehend and interact with the system through visualization of and controls through the ISM.

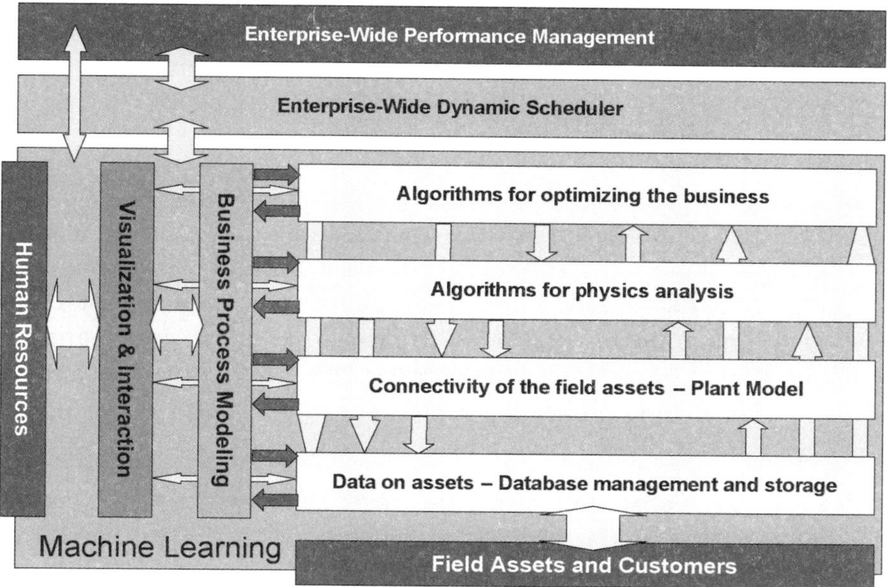

Fig. 3–1. The future vision of the ISM as viewed from an enterprise perspective. The ISM will take all incoming information about field assets and the customer and link it to a connectivity model that relates each asset with all other assets. The ISM uses the PM to compute its physics and business analyses of the assets, the BPM to improve the work being performed on the system by human and machine, and the ML system to mitigate risk.

In figure 3–1, we depict the building blocks that will lay the foundation for the ISM from an enterprise perspective. As with any lean business, it all starts with the customer. As a beginning, the customer's usage of your product or service needs to be accurately measured. In the case of electric utilities, automated meter data, or very accurate load research statistics that are based on actual real-time telemetry of electric meter readings, provide this point of origin. For oil and gas companies, consumers drive the market price to fluctuate as supply and demand for each refined product ebbs and flows. One little-known fact about the oil business is that a company does not refine its own petroleum products. Almost all produced oil and gas is sold on the trading floor, and refineries then buy their feedstock at best prices from sources geographically closest to them. The ISM can make the management of these complex consumer signals more efficient.

The need to supply customers with products and services requires deployed assets in the field. Data about these field assets are the next to be added to the ISM. This information is usually located in many different user databases, as with a pipeline; data on its size, capacity, and composition are likely kept in capital-purchasing databases, separate from its repair and inspection histories. The intent of the ISM is to integrate all of these disparate pieces of information in a model in such a way that any software algorithm can find and tap into this information and understand how it relates to the business of serving customers.

Assets are capable of providing revenue only if the customer is satisfied with the product or service delivered. Therefore, it is critically important that the connectivity of assets be modeled to properly describe the capability to develop and supply the product or service to the customer. This is true regardless whether refineries, oil and gas pipelines, power plants, the electric grid, or LNG trains are modeled. All have capacity constraints—in terms of volume (in amperes or gallons), pressure (in volts or pounds per square inch), time, price versus cost, and so forth—that must be modeled in order to fully optimize the performance of the business.

This initial ISM foundation creates the model of critical assets and their associations and interdependencies so that various algorithms that perform physical calculations, such as thermodynamics or electric power flow, can be computed. The same holds true for statistical algorithms that are critically important to correctly measuring business risks, such as reliability analyses, with which engineers are very familiar. An excellent example comes from the electric transmission business, where models containing up to 50,000 nodes calculate the reliability of high-voltage power supply to entire regions. Unfortunately, this model size limit does not come from the physical assets themselves, but instead from the capability of the matrix-based algorithms within the models that are presently used by the industry. These models still take an overnight computer run to get results—and they are relatively small compared to those contained in the Distributed Engineering Workstation (DEW) PM of the electric distribution network for Detroit, for instance, which we will soon see contains three million nodes but takes only minutes to solve.

In the future, the creation of a single enterprise-wide ISM—with its description of all critical assets, physics, statistics, and business economics and its ability to run modern simulations in seconds—will be achievable for all business processes that will include multimillion-node assets within the PM, to be discussed next.

Once necessary work is decided upon, there will need to be ways of efficiently issuing work orders and tracing effectiveness of the work within the ISM. Today, BPM software, as will be discussed subsequently, is available to coordinate and measure the effectiveness of the work. Also, business and operational intelligence algorithms in the form of ML, with which we will end this discussion, are available to perform *brutally empirical* analyses of the state of the system and ways to optimize the effectiveness of work.

Another very important use for the ISM will be to attempt to understand how people make decisions with skills that far exceed the powers of any current computer system, unless more than five variables are changing at once. In such cases, F-16 fighter pilots need computer assistance, for example.[3] The reliance on humans to make decisions will necessitate significant future improvements in the presentation of information from the ISM, to allow them to comprehend easily the state of the system and its resources. Visualization techniques have come a long way from the early days of SCADA. Above all, the ISM eventually will provide people with the capability to comprehend and make expedient and optimal decisions to effect positive changes to the system and business of the entire enterprise.

PLANT MODEL

ISM construction begins by building a PM that represents the physical layout and connectivity of the end-to-end process, whether it is an LNG train, an oil and gas production facility, an electric distribution grid, a power plant, a wind farm, or all of the preceding, connected into the entire business of the company. Critical components are then added at their locations (called *nodes*) within the topology of the PM. We hope that in the future, the ISM will connect PMs that begin at the source of the resource, connect through a power plant, refinery, or petrochemical facility of some kind to storage, and then to a trading floor where the commodity is sold and subsequently delivered. No model has yet been built that can connect these PMs.

The DEW model from EDD, Inc., is an open-source example of a PM for electric distribution systems—or other systems with flow in networks—and these can be combined to form a joint infrastructure model. One of the nicest features of DEW is that *objects* like transformers or compressors are placed in *containers*. Containers act as receptacles for objects. Objects can be swapped in and out of the containers, which allows DEW to easily change the components in a model as replacements and additions are made to the physical assets of the system.

Within DEW, each of the algorithms that compute the mass flow, heat transfer, or whatever the physical process requires to simulate the progression of the resource through the plant is solved in a self-contained way. This allows the algorithms to be composed together to form more complicated analyses. DEW has the best currently available chance of forming the core of the ISM that will solve the end-to-end enterprise-wide model.

Consider the example of electric reliability analysis. The PM algorithms have high dependency on one another, as shown in the boxes in figure 3–2. Common practice in the electricity industry is to perform reliability analysis without regard to voltage and current constraints, system protection considerations, or the ability to reconfigure the system via switching. Trying to incorporate all of these additional critical factors using legacy programming techniques would be extremely difficult. In the modern PM, algorithms are linked to dependencies on lower-level algorithms, and data are computed and flow as through a network of building blocks that can each be solved for a different, but complex, emergent issue within the PM (fig. 3–2).

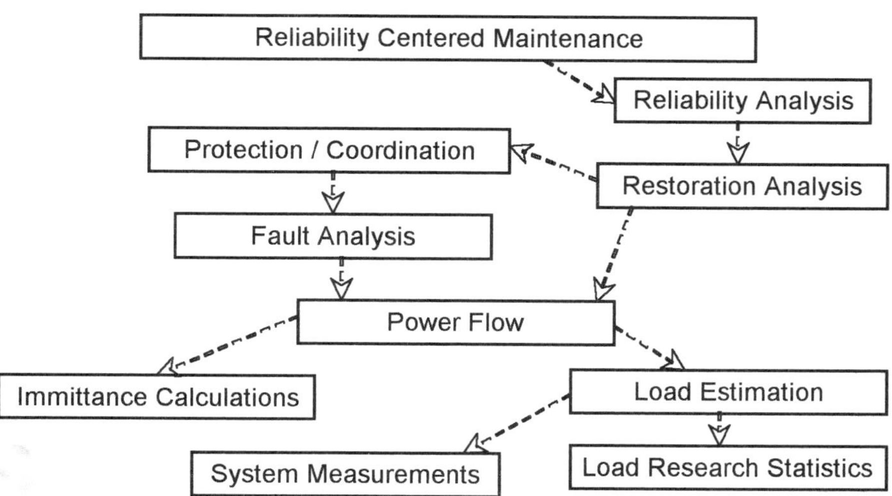

Fig. 3–2. The physical calculations to analyze electric system reliability within the PM. Ideally, all should be performed using the same PM, as in the DEW example.

Using the CALM approach of building a unified PM, each algorithm has access to the same data sets and results from all other algorithms attached to the model. This capability provides a collaborative environment that allows development groups from very different domains to work together through simulations and queries run on the same model. For example, the addition of electrical switches to the DEW system, which affects the ability to restore power, is managed by the restoration-analysis algorithm. The reliability-analysis algorithm dynamically links with the restoration-analysis algorithm to evaluate the full effect that adding a switch will have on both calculations. This empowers the developer of the reliability-analysis algorithm with the capability to leverage work and expertise from other developers.

Million-node PM

PMs of large grids like those in Detroit (fig. 3–3), St. Louis, and New York are being solved using DEW today. As far as we know, Detroit Edison (DTE) has the largest PM of any electrical distribution system in the world. It has been built using DEW and currently has more than three million nodes. A next-generation PM for Con Edison, in the long process of being constructed, will require more than five million nodes to describe the New York City electrical grid.

At DTE, DEW has been used for successful field control of seven different types of distributed-generation (DG) facilities that are used to prevent low voltages and overloads. Aggregate control of multiple DG sources has been demonstrated, whereby an automated interface to weather forecasts uses historical consumption measurements on circuits to forecast load for the next 24 hours. DG usage is then bid to the independent system operator (ISO), the market-maker for bulk electricity in the region. Looking to a future with much more extensive use of DG, DEW provides the functionality for designing its placement in circuits, including fault analysis and reliability analysis for DTE.

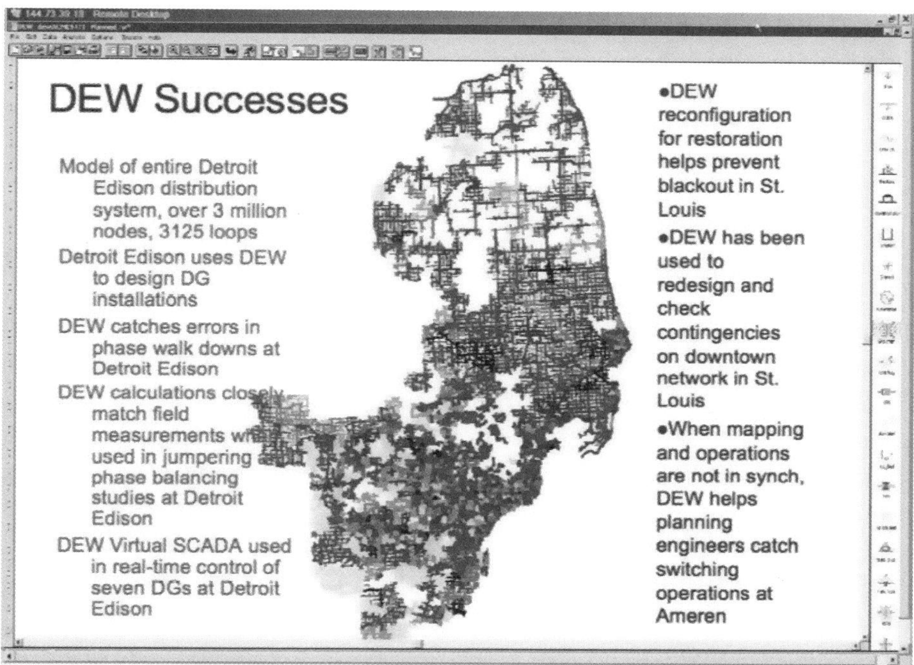

Fig. 3–3. PM created for the Detroit Edison electric grid, using the DEW software (EDD, Inc.) and results from Amgen in St. Louis (***Source:*** Robert Broadwater, EDD [http://www.edd-us.com/home.htm])

Immediate benefits from building the PM are consistently realized from harvesting low-hanging fruit—that is, improvements to system performance can immediately be seen once the PM begins computing simulations. For example, shortly after DEW was built, Amgen used it to reconfigure a large distribution network in downtown St. Louis to isolate a failing substation transformer, and customer power outages were prevented.

DEW was installed in its first operational control-center for the Orange and Rockland Utility of Con Edison to provide an integrated real-time analysis of operations. Control actions can then be prioritized using the PM, including such critical items as system recovery, survivability-based design, readiness/casualty response, preventive maintenance, and repair scheduling.

System interdependencies

Another feature of DEW's use in a PM is that system interdependencies (e.g., between an electric distribution system with gas-fired generators and a gas system with electrically powered control components) can be modeled together through the various algorithms that attach to the generic PM. System reconfigurations, physics-based constraints, discrete rule-based control actions, fully integrated system and component-level design analysis, control analysis, and data management can all be developed independently by different groups and can then be tested together in the collaborative PM environment. This is the beginning of the construction of an enterprise-wide ISM, though to our knowledge none has been entirely built out to date.

Infrastructure interdependencies

The PM of interrelated infrastructures can be connected across companies and jurisdictions to simulate interdependencies. For example, we are working to build a joint PM for the electrical systems of the New York City subways and the Con Edison electric grid. Resiliency for both systems can be quantitatively examined by simulating failures and asking the PM to calculate the effect on each. In addition, this approach to software integration allows for capabilities to model reconfiguration and recovery, equipment-scheduling interdependency, and infrastructure interdependencies—all at the same time. For example, gas distribution systems depend upon electricity supplied from the electric grid for powering pumps and compressors, and subways depend upon power conversion from the power grid (alternating current) to their third rails (direct current). In the future, we believe that the same PM that is used for design can also be used for real-time control of all interconnected systems simultaneously. Also, the PM can be used for any process—electricity, gas, oil, steam, sewage, or water flow—a critical functionality of the future ISM.

BUSINESS PROCESS MODELING

BPM is a methodology for mapping the as-is (and to plan the future) structure of the processes of the business, whatever they are. BPM is used to identify gaps and disconnects between *swim lanes,* or silos of responsibility during the flow of interactions in any process so that a to-be improvement in efficiency can be enacted through better cross-organizational integration. BPM mapping produces a flow diagram of how processes within a company are executed (see fig. 3–4). Many commercial and freely available open-source software solutions provide the mapping tools for BPM. The best use standards like business process modeling notation (BPMN) and business process execution language (BPEL). The ordering of the process flow can then be changed by dragging and dropping the boxes representing actions into different locations within the horizontal time-flow maps and reconnecting flow lines in a more efficient manner. The difference between BPM software and strictly visual mapping tools is that they know data flow requirements and software applications that reside at each location, so dragging and dropping an action and its flow path results in the BPM's reconnecting the required data flows into the applications, wherever they reside. Processes can actually be enabled for real operations by changing the connectivity within the BPM.

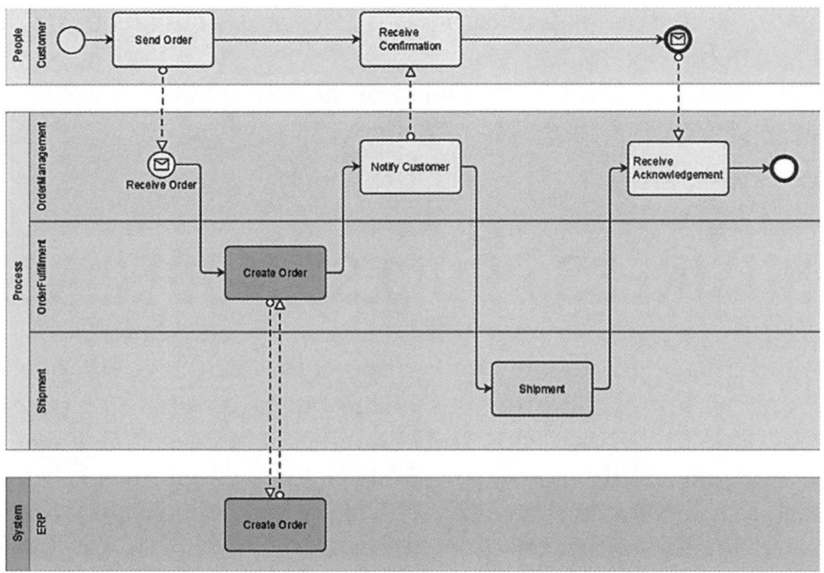

Fig. 3–4. Flow diagram based on BPM using open-source mapping software (available at http://www.Intalio.com)

Stages in BPM improvement are

- Identification of how a company functions, so that pockets of inefficiency, bottlenecks, and risks of providing service can be identified, mitigated, and corrected.
- Formalization of how a company should be operating with flowcharts, organizational structure, and performance metrics.
- Modeling of the processes of the company to actual roles, including technical details, exceptions, and flow of information.
- Conversion of processes to software code so that the computer can enforce improvements in efficiency.
- Deployment of that software so that processes can be executed as designed, to include interaction with legacy systems.
- Improvement in the performance of employees so that tasks will be driven by processes as designed.

- Monitoring and scoring of performance so that proper process improvement is executed.
- Analysis of variances of expected ranges, bottlenecks, suboptimized flows, and underperforming tasks.
- Optimization based on feedback loops that score whether actions taken result in beneficial outcomes.

The boxes displayed in the large area that takes up most of figure 3–4 represent tasks to be completed by people, assisted by processes run by the BPM software, or by some other IT system. In the upper left, the palette allows additional shapes, like tasks, to be created and linked onto the BPMN diagram. In figure 3–4, the model shows a customer sending an order and waiting for a confirmation. The BPMN shows a dashed line to the order-management lane of the process pool, which then automatically sends some type of confirmation acknowledgment (e.g., an e-mail) to the customer. The representation of communications across silos or pools (swim lanes) of different users is critically important to understanding and optimizing the drive against inefficiencies.

In figure 3–4, we see that the BPM process also takes the received message from the customer and performs a task called *create order.* That task communicates directly with the enterprise resource planning (ERP) system to create an order. The sender in turn receives an automatic communication back from the ERP to confirm the order. The process pool then performs the next task, called *notify customer,* and thus triggers a communication back to the customer in the people pool that will schedule the action contained within the customer order. The process continues running to its end point of fixing the problem after receiving scheduling confirmation from the customer. Underlying this action sequence, the BPMN creates executable code in BPEL to run the process to completion of the order and verification of satisfaction from the customer.

BPM enhances knowledge transfer and allows employees that are engaged daily in the execution of the processes to master their own organizational change. In addition, best practices are easily distributed to other groups by way of BPM. And critical to future energy delivery, intellectual capital is not lost when employees leave or retire if their knowledge has been captured within BPM.

In addition, BPM enhances interaction between business units and managers by enabling everyone to understand how the organization communicates to supply the customer with a product or service. Most important, BPM provides the vehicle for automation by use of software, and that, in turn, means measuring the success of actions while controlling costs more efficiently.

The introduction of BPM within a company is a differentiator in that the risks involved in its deployment are significantly lower than in the deployment of larger IT systems from mammoth organizations with proprietary software such as SAP or PeopleSoft. It solves many of the shortfalls of reengineering that have impeded SAP and PeopleSoft implementations in the past. In addition, once SAP or PeopleSoft committees agree on how change should happen, there is little capability of enforcing the change or measuring the success in terms of the value created without deploying extensive additional resources.

The risk inherent in purchasing such enterprise software from famous companies is low for the IT department's perspective. After all, this same software has been used in three hundred other companies already. The problem is that specific risks are huge for the company, which must now deal with *rigid* software codes (i.e., codes that are difficult to change and thus costly to modify). Continuous operational innovation requires flexible software.

Past experience in reengineering efforts like total quality management and capital process improvement have proved that in most cases, cultural resistance wins out or top management loses interest in the effort owing to the lack of tangible results. Thus, this approach rarely results in success and typically fails to make the leap to running the company with the kind of computer-aided rigor that CALM represents.

In contrast, the BPM methodology is to automate and economize the ways the company executes business processes in order to accomplish the reengineering in incremental steps that are driven by the true users. With BPM, a company can take one process at a time, manage change in low-risk steps lasting three to six months each, and extract value from this managed transition all along the way. In our experience, low-hanging fruit begins showing up after only a few BPM sessions via the creation of the as-is diagram of how messed up current processes are. The as-is diagram shows the true communication mismatches—and associated handoff challenges—between departments or silos. With this first as-is diagram alone, management makes small changes to the ways processes are performed, immediately benefits from measurable improvements, and wins over the people engaged in the change in the first place.

One real key to making BPM successful is the improvement of communications among silos that rarely speak the same language—namely, the business analysts and the IT department. BPMN enables these two very differently focused organizations to talk the same language and jointly develop software focused specifically on value creation within the enterprise. In addition, BPM software is flexible—that is, software changes can be quickly made as improvements in how to fulfill services to customers are identified. The capability to quickly and efficiently make continuous improvements in the business process is at the core of lean management, and its implementation is now possible with BPM.

Activity-based costing

BPM assists the organization in innovation by defining the business needs to be accomplished and then identifying the high-level requirements that must be fulfilled to get there. End-users are engaged in evaluating performance against business needs and in iterating and continuously improving the processes to develop the to-be state that the company is striving to reach.

The feedback loop that is required for successful BPM mapping consists of

- Plan:
  - Set objectives.
  - Define processes to meet them.
- Execute:
  - Track the actions.
  - Validate the quality of the work.
- Measure:
  - Monitor process execution.
  - Gauge performance against objectives.
- Correct:
  - Perform tasks to compensate for gaps and unexpected outcomes.
  - Learn from less than satisfactory outcomes and continuously improve.

Critical to the success of BPM is that each cost to do any task can be associated with that process. Both personnel and equipment costs can be accounted for. BPM enables the monitoring of process execution, to identify for each task the value added as well as the true costs. ROI can be quantified and efficiency identified.

To effect BPM in full, business activity monitoring (BAM) software is readily available as commercial off-the-shelf (COTS) software. Not having BAM to comprehend the BPM process is like have a fighter jet that can fly at Mach 3 but is missing a speedometer. Using BAM to manage process performance will be a key component of an integrated, enterprise-wide performance management system that eventually allows everyone in the company to comprehend the strategic, tactical, and operational performance metrics of the company and how their functions contribute to improvement in near real time.

Implementation of BPM with BAM

Identify a business process that will likely provide immediate benefits. A good approach is to use the methodology in Enterprise Value Stream Mapping and Analysis (EVSMA). MIT offers a free online course on the subject.

Create the change management plan and the as-needed cross-silo IPT. A plan with clear goals and objectives for coordinating and making needed changes is required. A team should be created that includes IT, R&D, and cross-silo department representatives.

Develop the as-is and to-be diagrams of the process. The IPT can then be empowered, through analysis of the as-is state and the development of the to-be process, to produce a truly effective process-reengineering plan that is specific to the particular problem within the company.

Implement change in the way you do business using BPM. It's critical to bridge the chasm between cultural change and organizational resistance by relying on the computer software to manage the required activities. Have supervisors ready to head off actions that do not reflect the needed changes.

Scheduling in an uncertain world

The value of having every stakeholder in the company on the same page is rarely realized because of the complexity of the supply system and the numbers of stakeholders, each interested in meeting their different operational goals within the company. Planning and scheduling equipment replacement is a major hurdle, especially for businesses that are already committed to supplying customers with reliable service on a 24/7 basis. Sacred redundancies of equipment are put to work to allow extra equipment to be taken out of service for repair or replacement. By coupling the capabilities for people management with BPM and asset management with PM, effective scheduling of work can be accomplished much more effectively so that these overbuilt redundancies can be minimized in the future—a major goal of CALM.

Figure 3–5 shows a CALM scheduling system where all can immediately comprehend the state of each workflow task—within the requirements for taking all sorts of equipment out of service, doing work on it, and putting it back in service. Each job can be related to work on other equipment connected to the system so that priorities can be set and reset as external events like emergency outages and weather perturb the system. Budget overruns, crew shortages, and capital equipment inventory shortages can be identified early and prevented. No CALM company should ever run out of any critical supplies or be short of critical personnel.

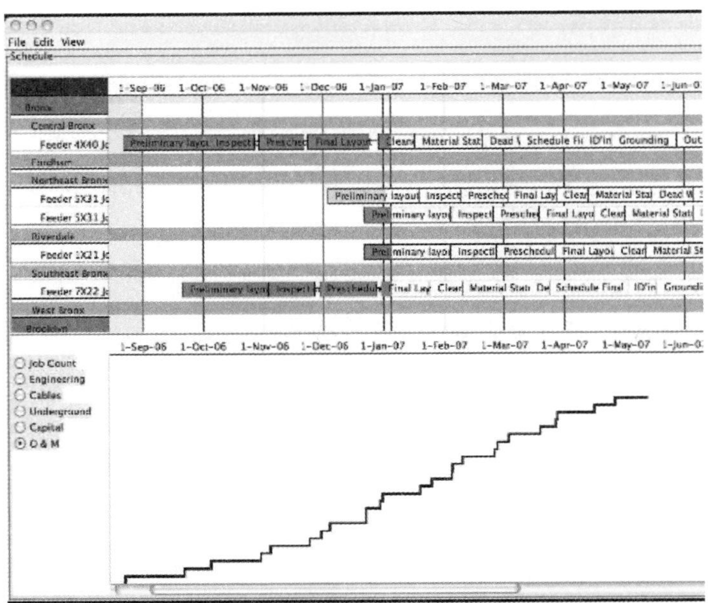

Fig. 3–5. Activity-based accounting, allowing specific costs and personnel assignments to be tracked for each workflow assignment

Dynamic scheduler

The visibility by everyone in the company of the scheduled work for each day should allow decision-makers to better coordinate specific work tasks with their associated resource allocations. Qualified employees can then be scheduled through the specific time frame within which the work needs to be performed. This resource allocation that is dynamic and responding to real-time changes within the system ensures that the right people and equipment are available to work on the right infrastructure at the right times. An offshoot of the dynamic scheduler's long-term planning for future work is that a detailed plan for training and hiring of employees to achieve these specific tasks can be made 6–12 months into the future. An example of an integrated dynamic scheduler for electrical operations is given in figure 3–6.

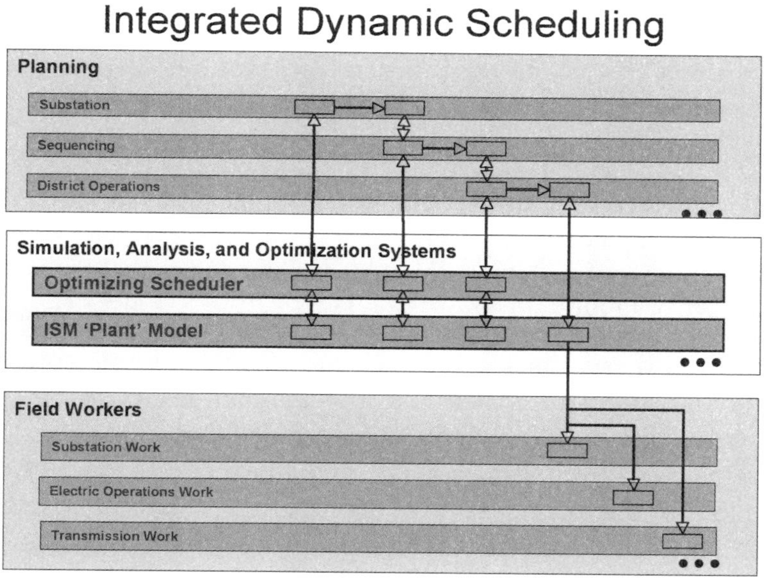

Integrated Dynamic Scheduling

Planning
Substation
Sequencing
District Operations

Simulation, Analysis, and Optimization Systems
Optimizing Scheduler
ISM 'Plant' Model

Field Workers
Substation Work
Electric Operations Work
Transmission Work

Fig. 3–6. Leveraging of the ISM and optimization algorithms to create a dynamic schedule of equipment and people that responds to changes occurring in the system in near real time

With the use of the ISM to coordinate schedules, costs, and resource allocations, portfolio management can be used to optimize work on the basis of simulations and risk analysis. Portfolio management involves selecting from a set of possible projects the subset that meets the constraints and maximizes the outcome while taking into account interactions among the projects. For example, a project planner is often faced with creating a portfolio of projects to repair, restore, or re-enforce over approximately a one-year planning horizon. Projects may include

- New construction
- New customers
- Reinforcement of existing infrastructure
- Maintenance of old infrastructure

Each project can be analyzed to produce value associated with

- Cost (capital and operations)
- Operational risk reduction (risk after vs. risk before)
- System vulnerability (risk during outages and system failures)
- Business risk exposure (risk in terms of the nonlinear relationship between per-customer cost of loss of service, the number of customers affected, and the duration of the outage or failure)

Schedule disruptions can affect costs, system vulnerability, and business risk exposure. Therefore, the planner needs to be able to produce schedules via portfolio management that deliver maximum improvement for minimum cost. The dynamic scheduler will be able to use the ISM in the future to derive quantitative business risk reduction for each project by comparing the mean time to failure before versus after the enhancement. We have built a capital asset prioritization tool (CAPT) that does just that for Con Edison (more on that in chapter 9). Even beyond that, the future dynamic scheduler may be able to use the PM to assess vulnerability risks during outages, while using a Monte

Carlo model to evaluate the time-of-repair exposure to increased business risk, so that schedules can be adjusted in the face of oncoming exogenous events like hurricanes or heat waves (fig. 3–7).

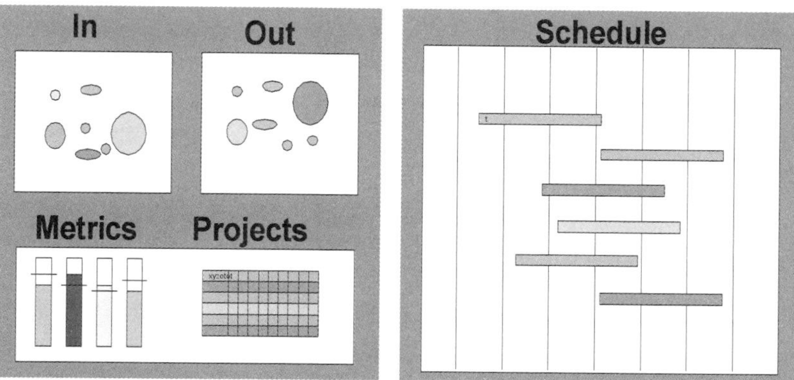

Fig. 3–7. Business risk reduction. The scheduler can drag and drop projects between the "in" and the "out" boxes and immediately receive instant feedback from the colored thermometer, where green means above acceptable constraints and red means cannot yet schedule the work.

Process mapping

BPM is used to assess the present state of business and to prescribe the courses of action that will lead to improvement. Some amount of process-mapping information already exists within any organization, having been captured (albeit in a less formal notation) during the design of workflow management systems (WMS). WMS typically employs Web-services architecture these days. In other words, WMS typically provides the software capabilities to store relevant information and provide communications of entries; which may or may not result in others taking action to move a process along. By contrast, BPM subsumes and monitors the process of getting the work done and verifies whether improvement resulted from that work. BPM thus provides the opportunity to incorporate existing WMS into an enterprise model that closes gaps, improves communications between the many silos of responsibility crossed during any project, and adds a dynamic-scheduling layer that sees to it that the work is done at the optimal time.

Tracking and coordination of the work effort is highly dependent on information about—and control of—all relevant processes, regardless of departmental responsibilities. BPM architecture is capable of the extensibility (i.e., creation of, encapsulation of, and interfacing with a new process) that is necessary to effectively ensure that process improvement truly occurs (fig. 3–8). Maintenance, inventory management, safety, and training of new personnel will all be improved if this CALM methodology can be established once the ISM has been built and the BPM is in place. No implementation of this combination of envisaged CALM tools and techniques has yet been built. However, at Con Edison, we have mapped in BPM all major operational processes related to maintenance activities carried out on electrical distribution feeders and their components.

In addition, BPM can be interpreted to establish metrics and goals for documenting improvements in cycle time and reductions in overall costs. Since it is likely that such improved interprocess coordination will be of highest value, identification and automation of the interaction points and flow of data and controls across process boundaries comprise the secret ingredients that BPM contributes to CALM improvement. Put BPM together with ML algorithms and the ISM will be able to create machine-aided continuous improvement throughout your business—or, in Toyota's words, "relentless pursuit of perfection."

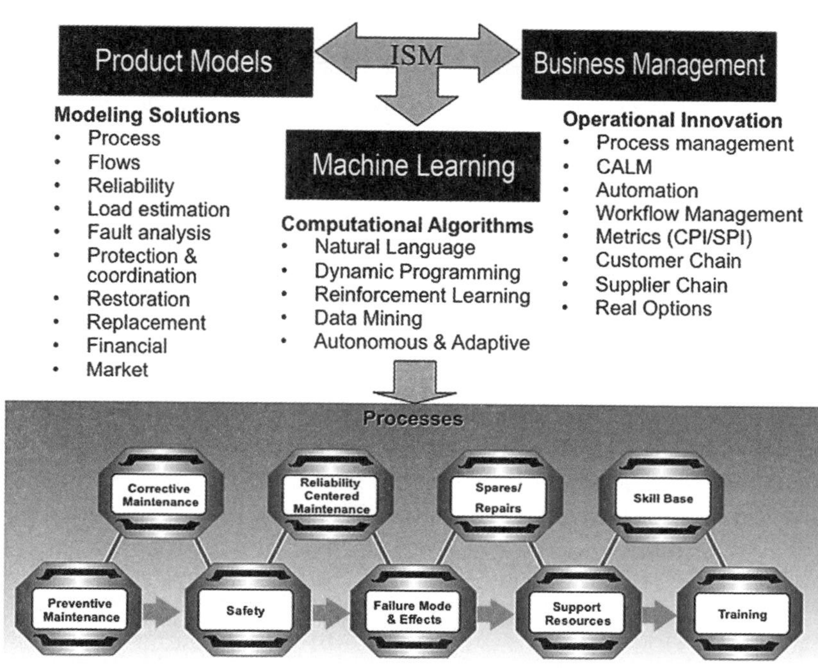

Fig. 3–8. Extensible BPM architecture for process improvement. The BPM, in combination with the ISM, computes, learns, and optimizes 24/7, so that processes that actually do the work of the company can be continuously and relentlessly improved.

PERFORMANCE MANAGEMENT

Modern business performance management (sadly, also abbreviated *BPM*) systems for corporations provide executives with strategy scorecards, or dashboards for how operational and budgeting metrics are performing day-to-day to achieve planned objectives. Although such systems have merits in steering the company from a strategic perspective, balanced scorecards in particular leave a lot of guesswork about how to achieve excellence in operations because they do not alone identify root causes for performance problems. However, if designed right, they can drill down to operations performance.

Another unknown is how well operations are aligned with the strategic goals and objectives set forth through the budgeting and planning process at the executive level. Because of extensive number juggling by departments that each may use different databases to justify their budget requests. Many of the key performance indicators at the corporation level have to be reconciled manually, even after the company has expended considerable time and cost to collect unbiased information. Drilldown into actual operations metrics is practically nonexistent for most corporate performance management systems. Yet, performance metrics somehow get developed for multiple business units of the company.

Other performance metrics are developed for monitoring business activities related to process-oriented operations. However, these key indicators provide little to no understanding of whether performance in one area of the company is in alignment with the overall strategic plan. Are all personnel steering in the same direction with the same emphasis and urgency? Does the operations supervisor understand how his or her metrics flow toward the goals and objectives of the company?

Given the need to simultaneously manage business performance at the strategic, tactical, and operations levels of the company, a single integrated software platform will better answer these overall performance questions. With intensive development in the future, CALM will be able to provide the foundation of the ISM to create an integrated execution strategy for the entire business. Integrated planning of the business will be a key element to unified steering of the company in the direction of increased profitability. In addition, corporate executives should be able to drill down into their operations to understand what the company is good at and to expose elements that are hindering or blocking strategic progress. If a company wants to grow from what it knows where it believes it has a competitive advantage, measurements of excellence in its own performance and operations should be readily available to confirm the validity of the proposed strategic direction (fig. 3–9).

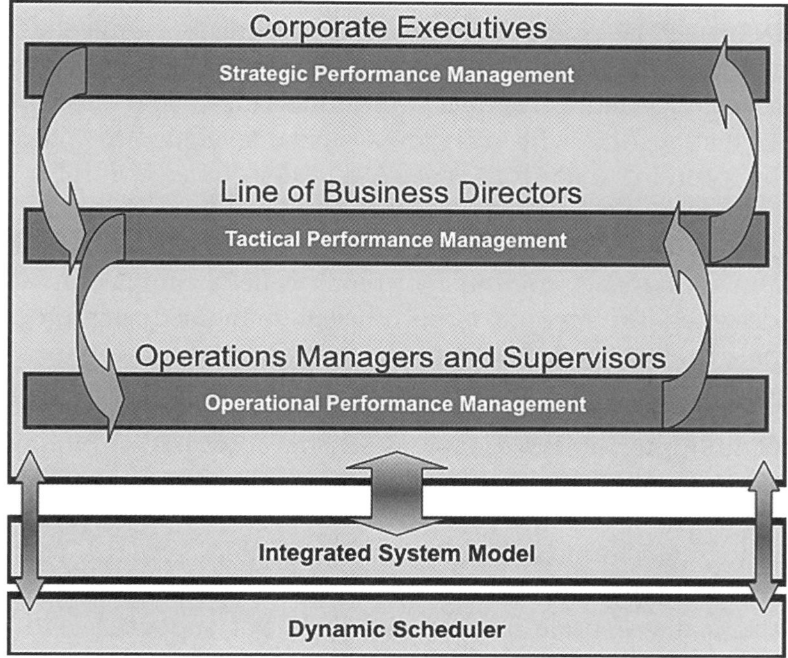

Fig. 3–9. The performance management platform. This should include the capability of simulating the business with the integrated use of business intelligence to uncover budgeting, planning, scheduling, and profitability opportunities and risks.

COMPUTATIONAL MACHINE LEARNING

From the 1970s through the 1980s, a new generation of manufacturing factories, making pharmaceuticals, semiconductors, cars, and airplanes, were built around the world. Control-centers of both new and existing facilities were modernized and brought into the Information Age, with software systems developed to control not only the incoming data from SCADA and distributed control systems (DCS) but also the processes on the assembly line. Data historians were created to store and archive the data from the SCADA and DCS sensor measurements coming in from the field. Simulation models were designed to compute more efficient manufacturing processes, and considerable effort was expended to push the most important information to the control consoles of each operator at the right time.

Beginning in the 1990s, process sensing information proliferated, driven by cheap sensors and computer memory, giant leaps in processing speed, and the explosive growth of the Internet. Control-centers migrated to more sophisticated computer support systems for operators. Not only was information on the state of the field operations delivered to the console in near real time, but support began to be provided about the most efficient process control decision to make at any given time and under any given condition. This progression in computer support was from *information management* to *knowledge management*. By the turn of the 21st century, such totally integrated enterprise management systems were installed in most control-centers of mega-manufacturing facilities in the world.

Most of the currently operating energy factories that utilize at least some lean management practices—such as refineries, petrochemical plants, and nuclear power plants—use processes that are centrally controlled. In these factories, operators monitor SCADA sensors distributed throughout the manufacturing facilities. These "wired" facilities have steadily evolved from process control and information management to computerized manufacturing management, plant optimization, and finally modern enterprise management within what Honeywell calls the *total plant* (fig. 3–10).

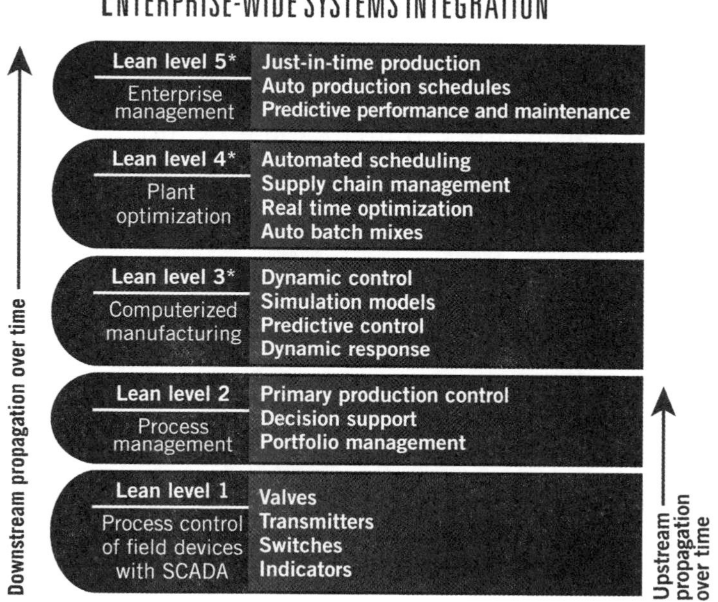

Fig. 3–10. Progression in lean management over time (up arrow). The downstream oil refinery business is at the top rung of this lean ladder, but upstream production is at about level 2. This lean ladder for the electricity industry would place nuclear power plants near level 3 and transmission and distribution control systems at level 1.

All energy companies already have some as-is legacy of process control and information monitoring that they are managing. CALM attempts to clean up conflicts between silos and integrates these with new tools for better knowledge management. For example, we have found that efficient workflow management requires knowledge, not just information, and that optimization can be accomplished only if lessons learned from measurements of the effectiveness of past work are used to improve future work. Therefore, CALM introduces computational machine learning (ML) tools and techniques to smarten the enterprise.

As we approach the close of the first decade of the 21st century, business decisions are working their way into the field. Knowledge systems are being augmented with ML that enables "drinking from the fire hose" of data streaming in from an increasingly sensed, measured, and monitored enterprise. Closing the loop with business-level push of information empowers control-centers to take actions that modify processes on the basis of the lessons learned from analysis of past performance, as well as current market conditions. ML allows a system to respond to shifting patterns and adapt via learning the emerging patterns. A car navigation system is one simple example of how such an adaptive system can respond to a change in the current situation. The car's Global Positioning System (GPS) adapts when you make a wrong turn and immediately recomputes a newly recommended course to your destination. This has been termed *adaptive aiding.* A goal of CALM is to provide adaptive aiding to all levels of an organization through the use of ML.

Artificial intelligence (AI) from the 1980s has grown up and is providing the computational learning tools that allow control-centers to improve their collective IQs by correcting inefficiencies dynamically. Today, the as-is technology on the "'net-centric battlefield" at NASA and within the aerospace industry is setting the standard for the to-be world of CALM for the energy industry.

Data mining has become a very popular tool for the enhancement of business intelligence. In general, online analytical processing (OLAP) data cubes are made of the mass of enterprise data residing in corporate databases, and data mining tools that incorporate scalable statistical and ML algorithms are used to ferret out patterns from the data.

CALM empowers continuous data analysis, modeling, performance evaluation, and ML to improve performance. The increased costs from migrating to this new, *sense-and-respond* operational framework are easily offset by subsequent decreases in capital expenses and operations and management costs produced by efficiency increases, as have been documented in industry after industry.

ML systems contain new and mostly unfamiliar computational learning aids, so we will step through the progression of technological complexity in more detail. We will begin with data mining and end with reinforcement learning (RL). Within RL, actions based on information coming in are objectively scored, and the metrics that quantify the effectiveness of those actions then provide the feedback loop that allows the computer to learn better actions in the future. In other words, once RL software is in place, the continual recycling between decisions and scoring of success or failure throughout the organization guides personnel to the best future actions. Only then can field industries that deal with daily variations in weather, traffic, cost and price swings, and other natural and man-made emergencies successfully adapt CALM technologies and techniques.

ML is a methodology for recognizing patterns in past actions by the analysis of data, rather than the intuition of engineers. Common algorithm types include

- *Supervised learning.* The ML algorithm generates a function that maps inputs to desired outputs. Given a set of examples (the learning set) of associated input/output pairs, a general rule, representing the underlying input/ouput relationship, is derived that may be used to explain the observed examples and predict output values for any new, previously unseen input.
- *Unsupervised learning.* The ML algorithm generates a model for a set of inputs. Similarities among examples or correlations among attributes are discovered (i.e., geometric clustering, but without known labels).

- *Transductive learning.* The ML algorithm uses small training samples to develop supervised clusters that allow labeling of the rest of the data (i.e., unsupervised, but similarly clustered data received in the future).
- *RL.* The ML algorithm learns how to take specific actions in order to maximize some long-term payoff or reward.

Data mining, to begin, is used to ferret out patterns from data. Figure 3–11 contains a simple example that uses data mining to discover general rules in the data by principal component analysis (PCA). Note the "manifold" pattern in the 3D plot of the first three principal components of the PCA, indicated by the arrows that sweep from the lower right to the lower left and then upward to the top center. This pattern recognized by the PCA indicates that at the time of this analysis (1998), there was a progression of oil company core competencies—from those that were cost conscious (Elf and Total), through those focused on reserves replacement (Anadarko and Apache), to those that dominated in profit through gigantic production volumes (Exxon, Mobil, and Shell).

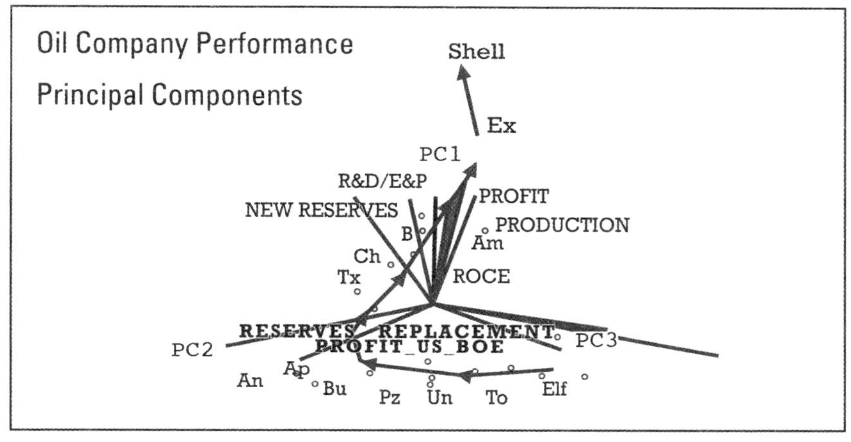

10 Database records

	$/bbl	$/bbl	%	%	%	Years	$ million	$ million	boe
Company	Costs	Earnings	CAPEX	ROCE	Reserves Replacement	Reserves Life	Net Income	EXP CAPEX	ONG Production
Amoco	14.50109	2.508907	0.589924	0.116119	1.17	9.4645089	1793	693	1366
ARCO	13.74627	3.749798	0.822711	0.20308	1.11	11.40255	1688	651	1029
BP	17.31787	3.1	0.905615	0.091	1.044623	13.67376	1704	823.4	1557.5172
Chevron	12.96729	2.854578	0.810826	0.12537	1.07	9.494709	2136	642	1446.3667
Elf	12.07798	3.389263	0.72805	0.115183	2.78	10.6739	1802.752	596.3303	732.96552
Exxon	15.51	4.294639	0.722535	0.129596	0.98	11.4422	5121	957	2694
Mobil	13.1857	2.648492	0.826999	0.098637	1.14	11.29302	1921	686	1516.5
Philips	9.69159	3.622322	0.708867	0.228101	1.823864	13.21236	1017	126	447
Royal Dutch	16.72639	3.021625	0.78589	0.104083	3.401813	14.60465	5523.94	1278.04	2942.2069
Texaco	12.43264	3.150704	0.923619	0.13874	1.201901	9.311686	1535	499	1172.1667

9 Attributes

Fig. 3–11. PCA of the 1990 performance of major oil companies (top). PCA classifies their performance statistics (bottom) into recognizable components.

Great opportunities exist for applying data-mining algorithms to optimize the planning process, so that investments are made in the right place at the right time, every time. Recognizing this manifold, Wall Street analysts could perhaps have understood some of the logic behind the wave of mega-mergers that swept through the oil industry in the late 1990s. Given the example in figure 3–11, could performance indicators from 1990 have been data mined to predict which companies acquired which years later?

Consider further analyses of the matrix of nine performance indicators for the top 10 major oil companies in 1998. A basic concept in data mining, as well as in ML, is to plot each row in high-dimensional space—one dimension per attribute—called the *feature space*. In the example in figure 3–11, that would result in a nine-dimensional space. People have difficulty comprehending more than four dimensions (e.g., length, width, depth, and time), so data mining and ML are of help. In figure 3–11 top, PCA was used to reduce the dimensions of this problem from nine to three. PCA recasts the dimensions into a new ordered set of dimensions, which were sorted by the degree to which each principal component of this new set explains the variability of the attributes.

We pick three traditional dimensions that define performance on Wall Street—earnings, return on common equity (ROCE), and annual income—to illustrate what a feature space looks like. This example will show a geometric view of ML methods that go beyond data-mining techniques like PCA. We can now place a small cube in this particular 3D space for each of the 10 companies. We label each with the name of the company for easy identification. We can now visualize what ML methods do in this feature space—including supervised and unsupervised learning and, to an extent, RL. However, RL has more than the feature space to visualize because it senses the environment in order to derive policies that recommend actions (more on that later).

Some ML methods, such as tree classifiers and support vector machines (SVM) (those with linear kernels, i.e.), place *hyperplanes* in feature space. (In figure 3–12, it is a plane, since we are illustrating a 3D example.) Such a hyperplane introduces a separating geometry. In supervised learning, the hyperplane represents the separation of classes of data that have been labeled ahead of time to be examples of a particular interest group—for example, symptom data from a database of patients who have a certain disease.

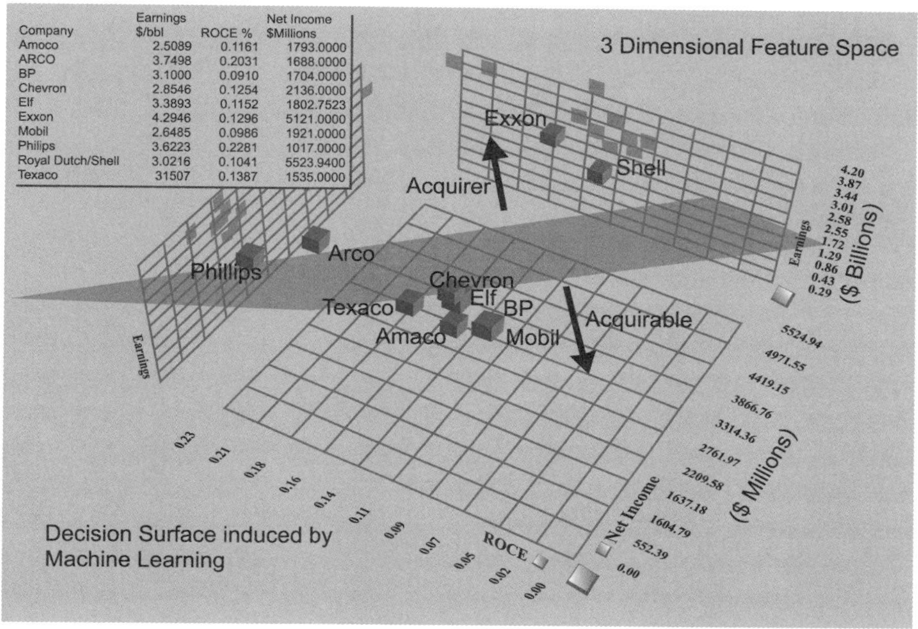

Company	Earnings $/bbl	ROCE %	Net Income $Millions
Amoco	2.5089	0.1161	1793.0000
ARCO	3.7498	0.2031	1688.0000
BP	3.1000	0.0910	1704.0000
Chevron	2.8546	0.1254	2136.0000
Elf	3.3893	0.1152	1802.7523
Exxon	4.2946	0.1296	5121.0000
Mobil	2.6485	0.0986	1921.0000
Philips	3.6223	0.2281	1017.0000
Royal Dutch/Shell	3.0216	0.1041	5523.9400
Texaco	31507	0.1387	1535.0000

Fig. 3–12. Hyperplane visualized. The plane represents an ML decision surface created to separate the earnings, return on capital, and net income of the oil companies.

In unsupervised learning, the hyperplane is deduced by the algorithm because the data are naturally clustering in a recognizable way without the algorithm's needing foreknowledge of what to classify each attribute. Supervised learning, by contrast, is good at identifying new syndromes or emerging problems because the ML algorithm is directed to treat well-separated clusters of data with the same label as groups. The problem gets more difficult when the data with different labels are intermingled or when a set of data with a label is physically separated from another set with the same label. Different ML algorithms induce different kinds of geometry into the feature space. Another clustering algorithm, the Gaussian mixture model introduces normal distributions, represented here as spheres around three distinct clusters of companies with similar attribute distributions (fig. 3–13).

Part of the practice of using ML is to decide when it is best to reinduce the learning of new geometries by reapplying the ML algorithms to training data and attributes. Breakouts from the historical clustering sometimes are not recognized quickly enough, causing investor losses on Wall Street, where many of these data-mining and ML algorithms are routinely used. How often the retraining occurs determines whether a firm will recognize these breakouts quickly enough to react successfully.

Returning to our oil company merger example, think of ML as discovering the geometry of aspects of the business capability or activity, in this case mergers and acquisitions (M/A). Dendrograms can then be used to visualize the relationship trees that lead from the individual company to the clusters of correlation determined by what are called *agglomerative hierarchical clustering algorithms* (fig. 3–14). The heights at which the branches in the tree join are determined from a similarity measure based on the specified attributes for the companies being compared.

We can peer into the boardrooms of these oil companies and try to understand some of the logic they used to direct their M/A decisions. In some cases, similar companies merged (e.g., Chevron and Texaco and ARCO, Amoco, and BP), and in others, acquiring companies have sought to supplement existing skills and portfolios by merging with dissimilar companies (e.g., Exxon and Mobil and, to a lesser extent, Conoco and Phillips) (fig. 3–14). Royal Dutch Shell has not found a match in recent years. Of course, M/A is also about less quantitative social and human interactions at the executive level, and this example is provided only to illustrate how ML capabilities might be used.

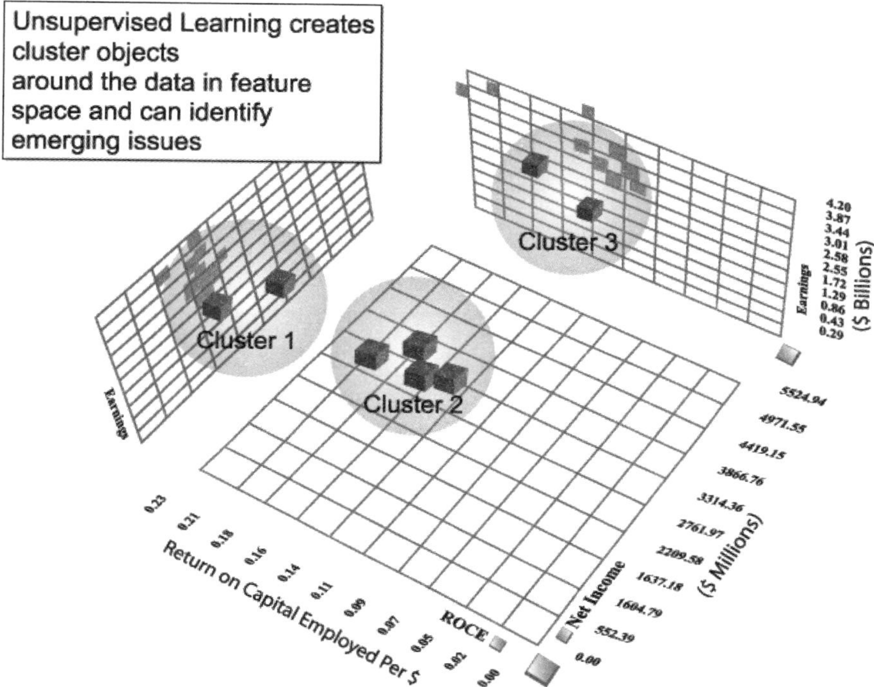

Fig. 3–13. Clustering, a form of unsupervised learning. The ML algorithm is allowed to recommend clusters that it detects, often without being able to identify why the companies cluster as they do; in such cases, management can identify emerging competitive threats, for example.

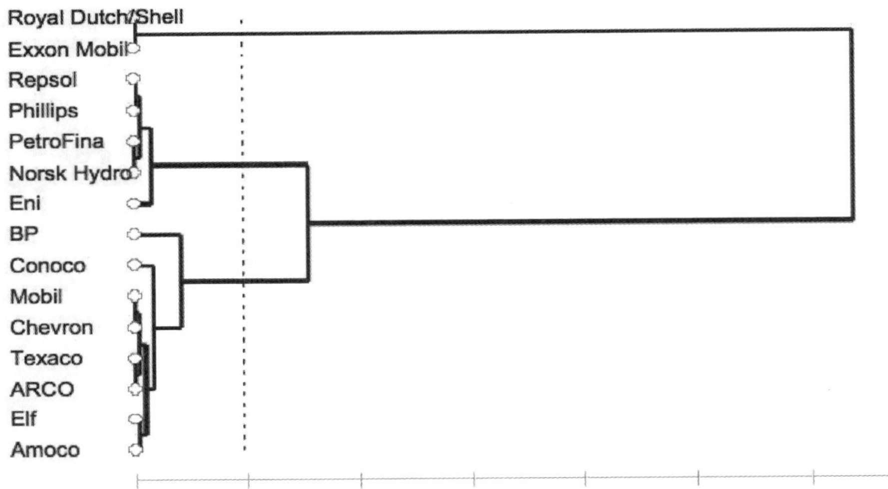

Fig. 3–14. A dendrogram created to relate the performance of the nine attributes for all 10 companies and determine which are most similar

Companies have recently extended the range of ML methods used for deriving predictions of future performance (fig. 3–15). Examples of successful ML usages abound. Google and Amazon use ML for searches and to tailor Web advertisements to user interests. In aerospace, ML has driven the progression from flight simulators that train pilots to computers that fly the plane completely on their own. Other successes include the progression from speech recognition to synthetic conversation and now to synthesizers of journalism itself. In the automotive industry, new navigational aids can actually parallel park a car, not just assist the driver.

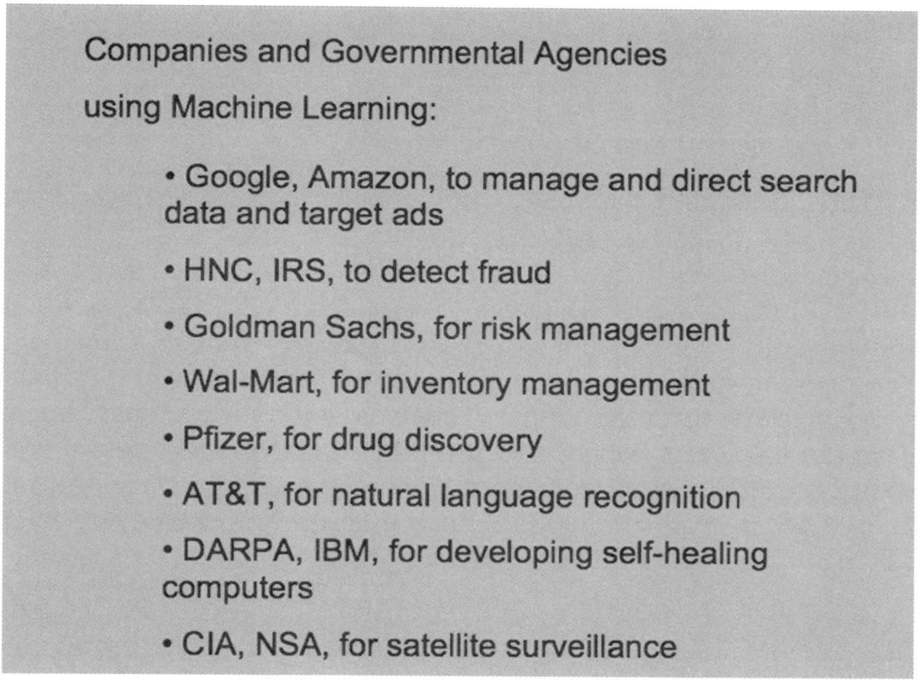

Companies and Governmental Agencies using Machine Learning:

- Google, Amazon, to manage and direct search data and target ads
- HNC, IRS, to detect fraud
- Goldman Sachs, for risk management
- Wal-Mart, for inventory management
- Pfizer, for drug discovery
- AT&T, for natural language recognition
- DARPA, IBM, for developing self-healing computers
- CIA, NSA, for satellite surveillance

Fig. 3–15. Varied uses of ML. ML has evolved from AI to its present dominant position in many of the most famous companies and institutions in the world.

ML methods effectively combine many sources of information to derive predictions of future outcomes from past performance. While, individually, each source may be only weakly associated with something that we want to predict, ML combines attributes to create the strongest aggregate predictor that it can recognize.

Suppose that we want to classify a data stream into like-performing characteristics. If we have a lot of data about what we want to predict, we will need a complex function that uses almost everything that we know about the object—and we will still have imperfect accuracy. We start with already-classified example data that we can use to *train* the ML system. Both ML and statistical data-mining methods allow the system to find good classifying and ranking functions in a reasonable amount of computer time, for even the largest of data sets. Although there are a wide variety of ML approaches, they have common features:

- Adding more data over time improves accuracy. With just a few data points, ML algorithms can make an educated guess. As the number of data points rises, the confidence and the precision of the predictions increase.
- Each ML technique prefers certain explanations of the data over others, all else being equal. This is called *inductive bias.* Different ML approaches have different biases.

The data that we wish to analyze will always have many *dimensions*—each of which is an attribute of the data. For instance, a compressor has a large number of attributes, such as its age, what it is used for, its load, and its configuration. Every object analyzed by a ML algorithm is described by a vector of these attributes. For example, compressor *x,y,z* has an age of 20 years, its location is in the 40s field, its peak load to rating is 80%, and so forth. There may be several hundred different attributes that describe the compressor and its past history.

High-dimensional mathematics works in a similar way to the mathematics of two and three dimensions, even though our visual intuition fails. The ML techniques described in this section have been extended to thousands—or even hundreds of thousands—of dimensions. Furthermore, special algorithmic techniques have been devised to make the larger-dimensionality problems computationally manageable.

In our example, any and all information about the compressor can be used to better determine the likelihood of when it will fail next—and, therefore, when it will need service. Real-time data documenting when the compressor turns on and off, in concert with vibration sensors, temperature, loading, and so forth, can be added as dynamic attributes to make real-time determinations of the compressor's continued capability to provide reliable service. This enables movement from preventive maintenance to predictive maintenance, based on the specific experience and existing health of the compressor compared to the history of other similar equipment.

Among the most basic problems considered by ML is learning how to predict which class an item belongs. The same techniques can subsequently be expanded from classification to rankings. For instance, ML can begin by classifying which compressors are at extreme risk, and then the data can be used again, to calculate a ranking of the risk of susceptibility to impending failure for every compressor in the inventory.

Neural networks (NN), as opposed to the newer ML methods, are still widely used in the energy industry. In current ML research and increasingly in applications, methods like SVM and Boosting have largely replaced earlier methods, such as artificial neural networks, which are based on a crude model of neurons.

These newer ML methods have significant advantages over earlier NN. For example, NN requires extensive engineering by hand—both in deciding the number and arrangement of neural units and in transforming the data to a form appropriate for inputting to the NN.

Unlike NN, both SVM and Boosting can deal efficiently with data inputs that have very large numbers of attributes. ML algorithms are not necessarily more complex than NN, but they have better theoretical foundations. Most important, these mathematical proofs *guarantee* that SVM and Boosting will work under well-defined, specific circumstances, whereas NN are dependent on initial conditions and can converge to solutions that are far from optimal.

SVM

The SVM algorithm was developed under a formal framework of ML called *statistical learning theory*. SVM looks at the whole data set and tries to determine the optimal placement of category *boundaries*, which create discrete classes with as wide a margin of separation as possible. In 3D space, such a boundary might look like a plane, splitting the volume into two parts. An example of a category might be whether a compressor is in imminent danger of failure. The SVM algorithm computes the precise location of the boundary—a hyperplane—in multi-dimensional space. SVM defines the location of the plane that best separates the attributes by focusing on the points nearest the boundary, which are called *support vectors.* SVM works even when the members of the classes do not form distinct clusters.

SVM algorithms produce estimations in which the training phase involves optimization of a convex cost function; hence, there are no local minima to complicate the learning process. Testing is based on model evaluations using the most informative patterns in the data (i.e., the support vectors). Performance is based on error-rate determination.

The discovery of a decision boundary between a grouping of case-history outcomes, with "go" decisions (light gray in fig. 3–16) and "no-go" decisions (dark gray in fig. 3–16), can be simplified by cutting the dimensionality of the data being mined. Thus, the decision space can be described by only a few support vectors that define the space—with error—between the light and dark gray clusters. The computer algorithm that defines the surface is the SVM.

Boosting

All companies struggle, given limited operations and maintenance budgets, to deploy enough information sensors to make accurate preventive maintenance predictions, so it is critical to distinguish those attributes that are really needed from those that are less useful. An ML technique called *Boosting* combines alternative ways of looking at data, by identifying views that complement one another. Boosting is based on the question, Can a set of weak learners create a single strong learner? Boosting algorithms combine a number of simpler classifiers that are based on narrower considerations into a highly accurate, unifying classifier. Boosting algorithms work by finding a collection of these simpler classifiers and combining them by use of *voting*. Each classifier votes its prediction and is assigned a weight by how well it does in improving classification performance, and the aggregate classification for an object is obtained by some voting rule. A weighted majority rule is common, but others have been used.

Support Vector Machine

Simplify the discovery of a decision boundary between a grouping of "go-decision" case-history outcomes in light gray versus "no--go" in dark gray by cuttting the dimensionality of the data being minded (whatever it is) so that the decision space can be described by only a few "support vectors: that define the space --with error -- between the dark and light gray clusters. The computer algorithm that defines the surface is the "Support Vector Machine"

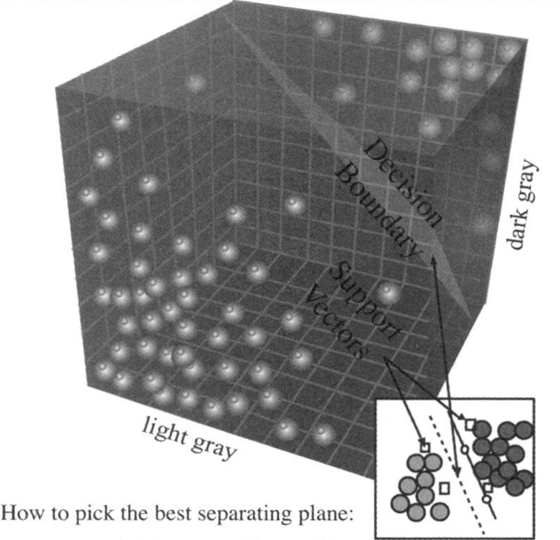

- Define a set of inequalities we want to satisfy
- Use advanced optimization methods such as linear programming to find satisfying solutions
- Statistically account for Noise
- Allow for Non-linear separating surfaces
- Surface is determined by the Support Vectors

Fig. 3–16. The SVM algorithm widely adopted among ML. The SVM operates as a black box to find the surfaces of separation (hyperplanes) among data in many dimensions that maximize the margin between the data categories. These boundaries allow predictions of future performance to be more reliable than traditional linear regression techniques or even earlier ML methods that did not maximize the margin. There are both linear and nonlinear versions (called ***kernel methods***) of the SVM algorithm. (***Source:*** J. Shawe-Taylor and N. Cristianini, ***An Introduction to Support Vector Machines and Other Kernel-Based Learning Methods,*** New York: Cambridge University Press, 2000)

Voting increases reliability. If each voter is right 75% of the time, a majority vote of three voters will be right 84% of the time; a majority vote of five voters will be right 90% of the time, a majority vote of seven voters will be right 92% of the time, and so on. However, this works only if the voters are independent in a statistical sense. If there are a *herd* of voters that tend to vote the same way, right or wrong, it will skew the results, roughly reducing the data to fewer voters. Boosting algorithms try to pick only a single representative voter from each herd—undemocratic, but effective.

In general, Boosting tries to select a group of voters based on two competing goals: choosing voters that are individually pretty good and choosing voters that independently complement one another as sources of evidence of the correct class of a typical object (fig. 3–17).

In CALM, we use the ML tools that best fit each problem. We try several, and usually one will work better on some portions of the data than the others. Our purpose in classifying is to gain fundamental understanding of the relationships between cause and effect among the many attributes contributing to the performance of each object. CALM can then manage many objects as integrated components of a larger system. Consider the failure prediction for compressors in figure 3–18. A good initial predictor can be derived by just looking at the number of past failures on each component, but nothing more is understood about cause and effect from an analysis of only this one variable. If, in addition, we apply both SVM and Boosting, we gain further insights into how the many attributes contribute to failure. SVM has an inductive bias toward weighting variables equally, and in this example, it does best at predicting future failures of the very most at-risk components (those ranked as the 10 worst). In contrast, the inductive bias of Boosting penalizes members of what appear to be herds; thus, it does best overall, and it predicts the 10 best compressors more accurately than does the SVM algorithm.

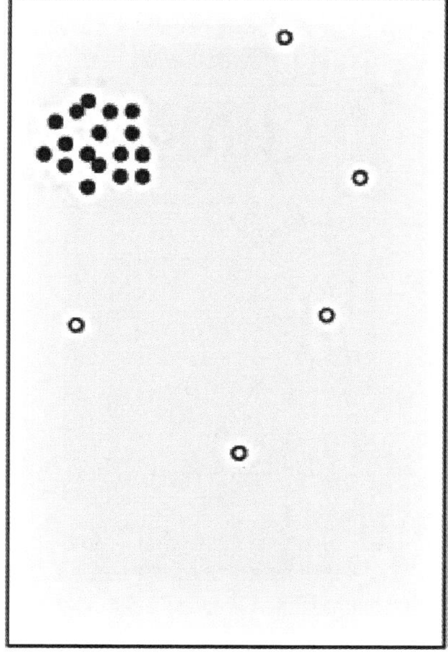

Fig. 3–17. Choosing voters. In Boosting, value is given to independence from clusters, as compared to SVM, where the hyperplane is used to assign rankings among large populations of objects with multiple attributes. The points at the left represent voters (not data), and the distance between voters corresponds to the extent of similarity in their behavior. (***Source:*** P. M. Long and R. A. Servedio, Martingale Boosting, 18th Annual Conference on Learning Theory, Bertinoro, Italy, 2005)

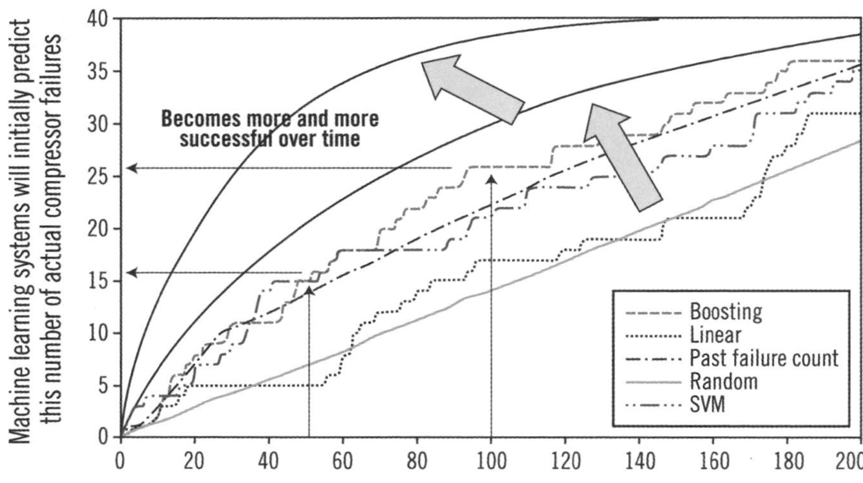

Fig. 3–18. ML ranking of susceptibility to impending failure of 200 components, such as compressors. Several ML algorithms are used to predict a ranking from most at risk (left) to least (right) so that preventive maintenance programs can take action to replace or repair those most susceptible to future failure.

While SVM often produces accurate predictions, it is difficult to get deeper intuition from them. In other words, SVM classifiers are more like black boxes than are Boosting algorithms. Insights may be possible if there are very few support vectors, but the number of support vectors rises rapidly with the amount of input data and the noisiness of the data, which is all the more reason to collect quality data. Boosting has a better chance of giving actionable results, and it is particularly good at identifying the hidden relevance of more subtle attributes—an important benefit when looking for what is important to fix in preventive maintenance programs.

The preceding discussion has described only static variables that do not change over time. ML analyses of the sequence of changes to dynamic variables over time provide an important additional determinant for prediction and root-cause analysis within CALM. Magnitude and rate of change of the variables can be used to derive the characteristics leading up to failure of a group of objects, so that precursors can be recognized for the overall system. For compressor failure, for example, accumulated cycles of load can be counted and added as an attribute. Such dynamic variables are important in fields such as aerospace, where airplane takeoff and landing cycles are more critical predictors of engine failure than hours in the air.

RL (approximate dynamic programming)

ML forces the quantification of operational and developmental learning into models. Outputs from SVM and Boosting algorithms can become inputs for a class of decision theory, control theory, and ML algorithms that seek to derive the best possible control policy. The result is an optimal sequence of decisions over time for a power plant or an oil and gas field. This class of algorithms, called *dynamic programming* (DP), can provide powerful decision support for operators and managers.

DP can become computationally intensive for problems that have more than a few dimensions of the problem because of the exponential growth of computation time with the number of states produced. This was termed the "curse of dimensionality" by Richard Bellman, who first developed DP in the early 1950s. A class of DP methods, known as *approximate dynamic programming* (ADP), make the computations needed to solve such problems doabl. Although RL was developed independently by AI researchers, in part to model the learning from the psychological theory of the same name, it has since been accepted that as a type of ADP.

Modern ML methods such as Boosting and SVM can be used as part of RL to help address the curse of dimensionality, although scaling RL to the number of inputs needed for real-time control is still a challenge. However, new methods to deal with thousands of variables are now offering hope for the future.

One of the earliest motivations to study RL was to understand how learning by trial and error occurs in animals. The interest was to discover how a simple signal could be used to train an organism to do complicated behavior by trial and error. Additionally, the concept emerged that the learning process could be sped up by computing many scenarios using a simulation model instead of using a real system to train a controller (fig. 3–19). In fact, some have postulated that dreams play a role like this in animals and humans. This simple yet powerful approach to learning endows RL with attractive features for practical implementation. It is a common approach in RL to build autonomous, intelligent software agents, experiment with their learning using the simulation model, and then use the agents for online operations. Although the real-time infrastructure for using this kind of RL for control is only now becoming available to industry, this architecture offers hope for adding intelligence to distributed, extended, last-mile systems.

RL and CALM make a natural pairing. RL lets an enterprise explore and evaluate its ever-changing options as they arise by using both real-time feedback loops and simulation models (fig. 3–19). RL is not told which actions to take, as with most forms of ML; instead, it discovers which actions yield the most reward, by sampling the possible choices. In the most interesting and challenging cases, actions will affect not only the immediate reward but also the next situation—and, beyond that, all subsequent rewards.

RL chooses these best operational actions by using techniques *very similar* to the ADP method of real-options valuation. In fact, one can construct the objective function for RL to be the expected payoff under an optimal exercising strategy—an optimal stopping problem.

REINFORCEMENT LEARNING FEEDBACK LOOPS

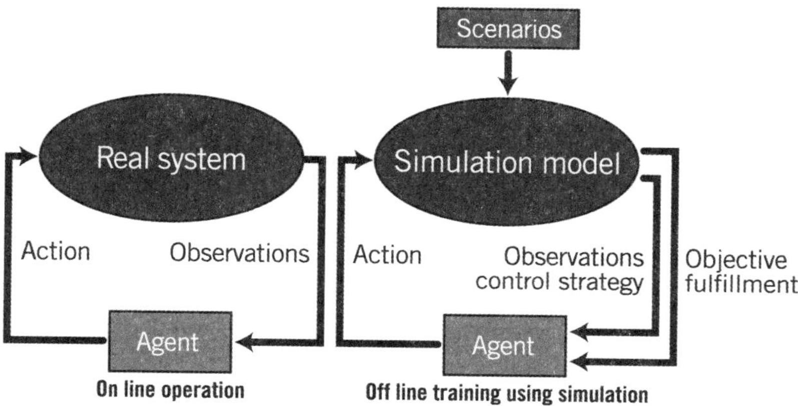

Fig. 3–19. A sped-up simulation to compute thousands of scenarios (right) versus a real system to train a controller (left). Simulation is a key component of practical implementations of RL schemes.

Revisiting the earliest motivation to understand how learning by trial and error occurs in animals, we introduce the following conjecture to be further explored in chapter 6. Critical to future progress of CALM will likely be the development of a feedback scheme analogous to the human body's motor control skills (fig. 3–20). Plant assets can be considered to correspond to muscles, bones, and the way the body dynamically relates to the environment. For example, the feedback of touch corresponds to the output of SCADA systems that are input to the control-center, the nervous system. Like the plant controllers, your nervous system needs to learn in order to better control its system. Through excellent transmittal via evolution of motor learning of how to control your body, the young can gracefully move around and, particularly in animals, quickly respond to threats.

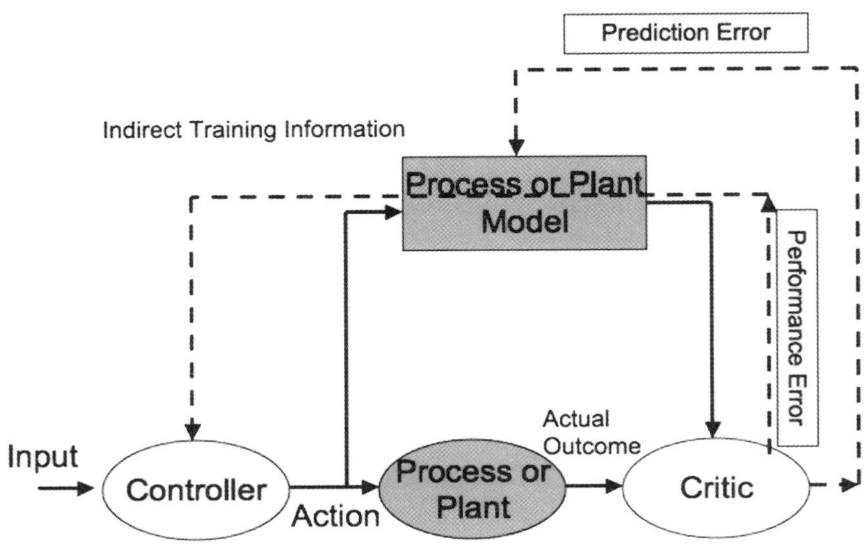

Fig. 3–20. RL as the process of taking controller actions on a real process or plant (the asset) and comparing the actual outcomes of actions within the plant to the actions that the controller expected (***Source:*** V. Gullapalli, Reinforcement learning and its application to control, PhD thesis, University of Massachusetts, Amherst, 1992 [http://citeseer.comp.nus.edu.sg/300268.html])

Referring to figure 3–20, in RL each action is rewarded or penalized using performance metrics determined by the critic. Dashed lines represent the paths that return the feedback errors from the critic to both the controller and the PM for continuous improvement. There is a method to derive the indirect training information called back-propagation that has been also applied widely to training NN.

A good beginning point for energy companies is to construct an accurate PM. With the PM, much more information regarding the real process or plant and its dynamics is available to the RL controller (fig 3–20). Moreover, the PM provides the means of incorporating domain knowledge that might be useful to the controller during bursts of uncertainty. Differentiation of real variability from noisy inputs makes construction of an adequate model purely through training difficult if not impossible, however. Fast forward modeling via ML tools such as SVMs can be substituted for the PM, to enable fast approximations of the plant's response to new inputs to address scaling issues that have plagued the RL's DP engine in the past (the curse of dimensionality).

Traditional PMs that use physics and chemistry equations to simulate the plant's performance can be used during idle time to generate training exercises (e.g., security threats), so that the system can improve without engaging the plant itself. This approach has been used in an RL computer game-playing system called TD-Gammon, to achieve expert-level performance playing backgammon by repeatedly forcing it to play against itself. This use of feedback loops is how companies can move closer to the goal of perfection in performance through continuous improvement contained within the CALM vision.

REAL OPTIONS

On the capital investment and long-term strategy side of decision support, another ADP tool, *real options* (RO), adds value to uncertainty by providing choices over time. The real option value is defined as the expected payoff in an optimal stopping problem within a control framework where the expectation is taken with respect to a risk-neutral distribution. RO methods are superior to the traditional net present value (NPV) technique of discounting into the future. In the energy industry, RO are excellent for computing added value arising from the nature of supply choices, the nature and timing of demand variations, the capacity flexibility available to transmission grids and pipelines, and the specific regulatory characteristics of the market.

RO bring the discipline of the financial markets to strategic capital investment decisions. RO calculate a value for investing a small amount of money now to keep an option available for executing a larger-scale decision, often a capital investments in the future. RO in the energy industry are fast becoming essential in budgeting for the construction of any large capital investment, such as a power plant, an LNG terminal, or a deepwater oil production platform. There must be variability in prices or markets into the future to give RO their value *beyond* the NPV—basically to be in a position to actually exercise the option. RO come in various types,[4] some of which are

- Input mix options or process flexibility
- Output mix options or product flexibility
- Abandonment or termination options
- Temporary-stop or shutdown options
- Intensity or operating scale options
- Option to expand
- Option to contract
- Option to expand or contract (switching option)
- Initiation or deferment options
- Sequencing options

Each of these applications of RO can be represented in a decision tree showing investments needed, expected timing, operating costs, technical risks, and benefits derived. Finally, the value of the RO is calculated taking into account the cash flows created from price variability, costs, and timing of operations for the various applications under consideration. Additionally, both the upside potential and the downside risk of each real option can be calculated and represented. A differentiator to the RO evaluation approach within CALM will be the ability to eventually simulate the system by using the ISM, driven by engineering, environmental, and financial uncertainties and allowing optimal investments and operating decisions.

Our approach, yet to be fully implemented, is to use the same ADP algorithm to optimize operations and compute the RO value. That way, not only economic interactions but also engineering and environmental interactions can be incorporated and policies enacted to avoid downside outcomes. This realization that the same ADP algorithm can simultaneously solve both the optimal policies and actions that a business takes in its day-to-day operations and the RO value of the flexibility of its choices in the future will perhaps make the automation of businesses possible. The combination of both real-time dynamic RO valuation and operational control would be a new business paradigm within CALM, but we emphasize that the vision has not been fully realized to date.

If successful, a portfolio of RO—and, therefore, the company itself—could be understood as a single compound, real option.[5]

NOTES

[1] Brownlee, S. 2007. *Overtreated.* New York: Bloomsbury.

[2] See http://en.wikipedia.org/wiki/Eastern_Air_Lines_Flight_401.

[3] Osinga, F. P. B. 2006. *Science, Strategy and War: The Strategic Theory of John Boyd.* New York: Routledge.

[4] Copeland, T., T. Koller, and J. Murrin. 2005. *Valuation: Measuring and Managing the Value of Companies.* 4th ed. New York: John Wiley & Sons.

[5] Cf. Brosch, R. 2001. Portfolio-aspects in real options management. Johann Wolfgang Goethe-Universität, Frankfurt am Main, Fachbereich Wirtschaftswissenschaften, Working Paper Series: Finance & Accounting, No. 66.

4 SYSTEMS ENGINEERING

Systems engineering (SE) has been the bedrock for managing new, complex systems in the manufacturing industries since the 1990s. It has been used for the complete life of assets, from the early exploration of new products and services right through to their delivery, operations, and ultimate disposal. SE enables the bridging of gaps between traditional engineering disciplines such as business analysis and software development and so has a natural fit into CALM. For example, training to work in teams to create products and services that deal with applications of complex technologies is a hallmark of SE. Complex projects require tight risk management, which involves taking calculated risks and managing them successfully.

On the oil and gas side, the wildcat-exploration tradition has reveled in taking huge risks, whereas on the electricity side, regulatory disincentives and perceived exposure to potential catastrophic incidents cause these companies to be averse to risk.

New game-changing technologies are now available that allow those companies to master developing complex systems while managing the risk.

With automation and software increasingly incorporated throughout any business, companies are searching for an integrated way of dealing with the complexity of the transfer of information to knowledge. SE is fast becoming a necessity in this regard. SE disciplines are required for successful implementation of CALM, to fully transform the company and its business. It is nearly impossible to accomplish this transformation without using the best practices of SE.

A goal of CALM is to provide a methodology and software toolsets that seek to integrate all designing, building, and operating processes of the company into a single openly shared, enterprise-wide ISM, using best practices like SE. CALM addresses planning, construction, installation, and maintenance—all together and inside the same ISM. However, uncertainties unique to the energy industry's economic evaluation, appraisal, fabrication, installation, and operation methodologies have to be integrated into SE if we are to succeed in transferring lean efficiencies to bottom-line profitability. Such a paradigm shift should immediately have a positive effect on performance.

The initial objective of CALM is to exploit the potential for significant improvement in cost and cycle-time savings by addressing the integration of operations and maintenance practices. CALM seeks to eliminate *white space,* or idle time in the workflow process. When coordination between operational silos fails, parts are not delivered on time, assets are not efficiently deployed, and management responsibilities are not effectively executed (fig. 4–1). SE addresses these issues.

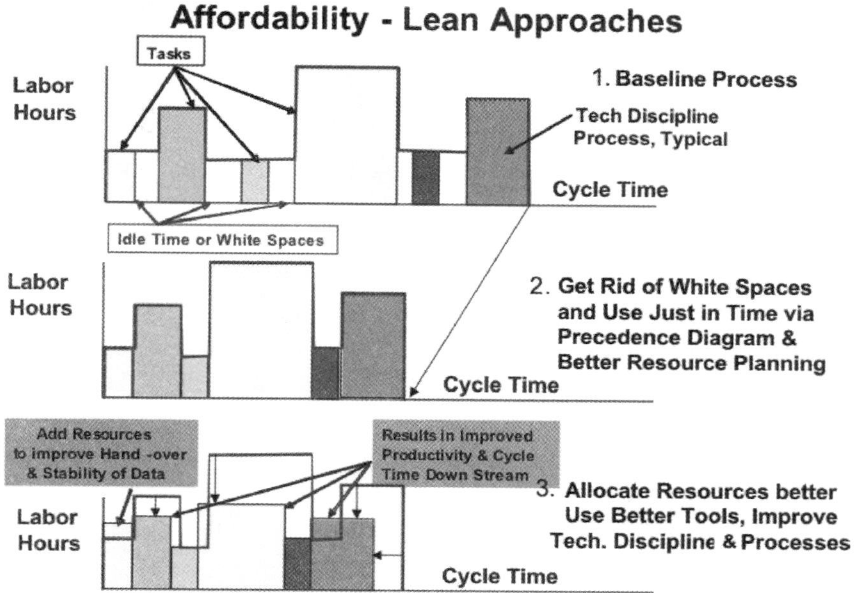

Fig. 4–1. Cycle time and savings of labor required for each project, as lean management focuses for manufacturing industries (from top to middle). Additional savings and profitability in field-intensive industries come from better allocation of physical resources and communication among human resources involved in each project (bottom).

If you are building from a baseline of your company's as-is state (fig. 4–1, top), the standard labor hours required for technical disciplines, such as operations and maintenance, are almost always accompanied by idle time lost while the organization is doing nothing to achieve positive goals. Much white space can be quickly eliminated by the mapping of the processes using BPM; this mapping step identifies the locations and causes of the disconnected processes responsible for the white space. A precedence diagram can then be created to establish, for all to see, who and what needs to be where, when to accomplish the tasks of the project. Implementation of this resource planning gives immediate savings, through the elimination of idle time (fig. 4–1, middle). In the oil industry, *time to first oil* is a critical measure of a project's success because, until then, capital has been spent with nothing to show for it. After first oil, revenues begin pouring in (literally), and cash-flow risk begins to quickly subside.

No labor savings have been realized yet, however. For that, more analysis is needed to improve the handover of tasks from one person to the next, and to identify where additional resources need to be added. If we were constructing new homes, for example, better coordination between framer, bricklayer, plumber, and electrician results in faster construction overall. Ironically, money must often be spent to improve communications so that fewer labor hours are wasted saving more than was invested in reduced idle time in the end (fig. 4–1, bottom). We might buy Web-enabled cell phones for our subcontractors, to enhance their ability to log onto the construction schedule at any time and track when they are needed on-site. In the end, CALM enables an improved to-be state of all projects by integrating

- Lean tools for tracking processes through the life of each project—from appraisal, through planning, construction, operations, and all the way to abandonment
- Lean processes for planning, scheduling, supervisory control, regulation, management, security, and environmental improvement
- Included in CALM integration methodologies are major cost reductions and operability gains from enhanced project visibility and transparency using SE techniques. Increased innovation and the sharing of knowledge follow, and better understanding of the goals and objectives of each project are attained throughout the enterprise. These gains are realized only if the vast majority of the people of the enterprise adopt lean management practices and break from the past to make relentless pursuit of perfection possible.

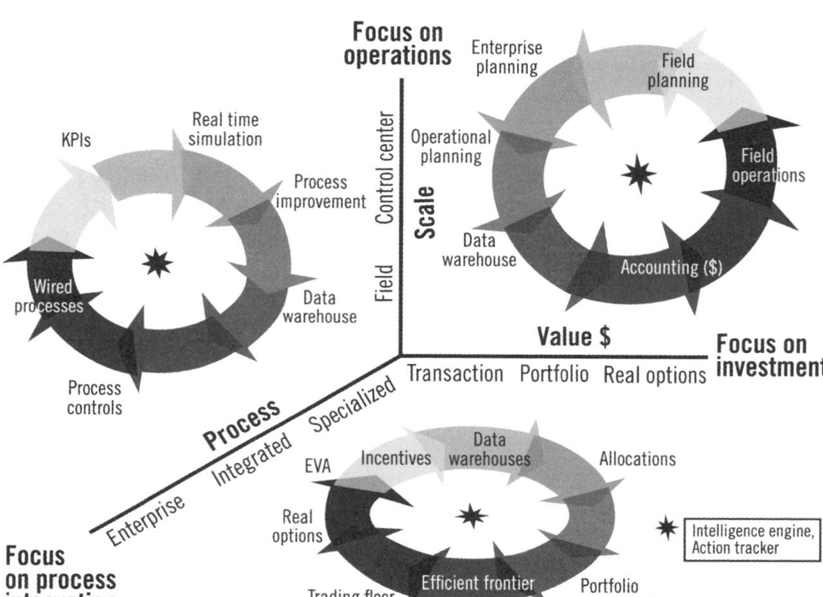

Fig. 4–2. Three complementary dimensions coordinated for the business to function smoothly. Disconnects between process improvement software tools, metrics, and simulations (top left plane) can slow planning and financing of operations (top right plane), and problems with both can severely hamper allocations of resources to maximize profitability (bottom plane).

Three complex axes of knowledge contribute to profitability in any business (fig. 4–2):

- *Process integration*—tracking the performance of all important tasks of the enterprise and comparing the actual with the planned and simulated performance.
- *Operations*—continually turning performance signals and metrics into day-to-day execution of the business plan.
- *Investment*—proper allocation of capital, equipment, and human resources, supporting both the execution of all processes and the operation of assets of the company.

These three dimensions of the company need to work smoothly together. The ISM provides the intelligence engine and the action tracker needed to make these three knowledge axes work in unison, and SE provides the know-how.

In field-intensive industries dealing with costly infrastructure—often scattered all over the world—it is difficult to make this machine run smoothly. A sophisticated intelligence engine is needed to identify and drive out inefficiencies and reduce business risk. Not only is ROI expected, but profit minus risk-adjusted cost of capital, or economic value added (EVA), should be maximized for the company to be most successful.

More accurate analyses, better accounting for uncertainties, and elimination of waste in the business are the results. The change within any organization from SE results in a major redefinition of IT duties, in particular. No longer is ownership of data or computer processes confined to the IT department. Instead, SE distributes data and software outward, to the locations of the work itself. Alpha and beta tests of innovative codes are encouraged. Confinement to many-years-old software products that hinder collaboration is discouraged. The use of open-source and COTS software is encouraged when they fit the requirements of the enterprise. Applications of cheap, distributed technologies such as those that improve SCADA system deployment are encouraged.

Another critical CALM paradigm shift affects IT departments: integration of subsystems is enforced by software. When SE changes a design specification, it propagates instantly to all other users of the software system. Earlier supervisory involvements result in fewer inspections—and many fewer parts, outages, and, in particular, rework orders. Cost and schedule controls are transparent and, therefore, available for all to see, at all times.

A key to the success of CALM is the simulation of requirements and appraisal of needs by using real options for quantifying cost versus the value of each addition. Analyzing costs and defining the structures to be built or repaired, then planning the execution of the chosen designs, testing operations, and evaluating the options required by customer and markets under uncertainty—all are done on the computer before the first piece of wire or pipe is laid.

The reason that real options are the preferred financial analysis choice over NPV is that they require flexible operating principles to be applied to the "factory"—whatever it is. Real options are the modern way to quantitatively evaluate the costs and benefits of adding this flexibility.

Another primary function of CALM is to simulate the possible outcomes from many options using the ISM. Only if the system can be adequately modeled can we fully evaluate consequences before we take actions. Battlefield concepts—such as situational awareness, global visibility of assets and inventories, distribution and transportation logistics, and optimization of supply chains—then become important to the energy industry.

Common actuators that SE brings are single source of product, standardization across platforms, parts grouped into assembly kits, and, above all else, large-scale systems integration. Resulting improvements are a digital library that allows attributes and geometries of all parts to be matched, to create standard configurations and suppliers. Reduced design, tooling, and manufacturing instructions increase procurement efficiency, lower manufacturing lot sizes, minimize artificial shortages and surpluses in parts, and support just-in-time delivery. In the following section, we examine the SE processes that directly enable CALM and discuss the tools and processes required in order to enact each in the energy industry.

SE COMPONENTS

Paradoxically, CALM shortens cycle time and cost by maximizing computational time. Preserving real options is essential should later understanding of uncertainty yield a revised set of operational parameters as each project matures. This dichotomy of shortening cycle time while providing more decision time is a hallmark of lean management processes.

SE components eliminate waste and miscommunication, while ensuring seamless information flow downward and among the owners and the contractors involved in any job. The SE approaches as they relate to one another are as follows:

- *Product life-cycle management* (PLCM) is the enabler for the integration.
- *Engineering integration* across disciplines leads to better designs.
- *Feature-based design* is used for parametric modeling, morphing, and standardization-driven supportability.
- *Virtual supportability* visualizes the environment and the workflow and generates electronic work instructions.
- *Supportability plans* are mapped out on the computer before any actual operations and maintenance begin.

PLCM

PLCM seeks to link all data and processes across the appraisal, selection, definition, execution, and operational stages of all projects. The result is faster, quicker, and better performance through enforced discipline, transparency, and continuity of design and requirements over the full life cycle of each project. PLCM consists of a set of software tools that enforce this process rigor across the stakeholders all day, every day, throughout the life of each project.

The following key processes are driven by the PLCM software tools:

- Applies simulation techniques to predict system behavior.
- Uses best practices and lessons learned to improve future performance.
- Uses a paperless, totally digital, design-anywhere/build-anywhere concept.
- Identifies specific owner-supplier relationships, including incentive.
- Performs earliest possible studies of alternative responses to uncertain outcomes.

PLCM improves capital efficiency by

- Shortening time to "go/no-go" decisions.
- Allowing fewer changes that affect costs and schedule.
- More effectively using internal and external suppliers.
- Improving understanding of asset management.
- Delivering automated performance metrics 24/7.

In SE, everything works off the same model of the entire system being built. Automatic data updates ensure linkages between analysis, simulations, and ultimate designs, shortening the definition and testing cycle. Smart software flags interferences among bulkheads, electrical wiring, and piping. A digital parts library is integrated into the conceptual design process to limit the number of components and increase commonality among projects. Requirements are clearly recorded for all to see.

Risk management tools are used throughout to identify, assess, categorize, and strategize through a formalized risk mitigation process. An integrated planning process creates precedence diagrams for all subsystems and then computes a resource load schedule. An action tracker provides daily progress metrics to all, indicating actual costs and whether all subassemblies are ahead of or behind schedule.

Assembly is planned on the computer before any actual physical work begins. A BPM map is created from start to finish of the project that is signed by all. Budgets are prepared from this bottom-up process map. All this is done with open-source or COTS software to ensure availability throughout the chain of responsibility, including all subcontractors. Program evaluation and review technique (PERT) charts and electronic work instructions are then issued, and task schedules are calculated so that the project can be staffed. PLCM provides a single source for all procurements and bills of material.

The PLCM for manufacture of an airplane is similar to that in an oil field, a refinery, a power plant, or an electric distribution system, except that uncertainties in the energy industry are much harder to define. As the sequential phases of the project proceed, PLCM enforces coordination and communications with software and Web services. Every stage is tracked, scored, measured, and evaluated for course corrections that might be needed because of uncertainties in the marketplace.

The PLCM of the Boeing 787 provides an excellent example of the benefits and hazards from the maintenance of real options. The wheel (fig. 4–3) provides for modifications to the design of the aircraft as late into the building process as feasible. The Boeing 787 was initially commissioned in 1999 as a replacement for its aging fleet of 757 and 767 aircraft. The price of oil back then was $15 per barrel and lower. The airplane was designed to be the quietest, most comfortable, and most fuel efficient in the air. It came under immediate fire by Wall Street when Airbus almost simultaneously announced the world's largest airplane, the double-decker A380. Which would sell better to the global airline industry—quiet and comfortable economy or a massive increase in revenue passenger miles?

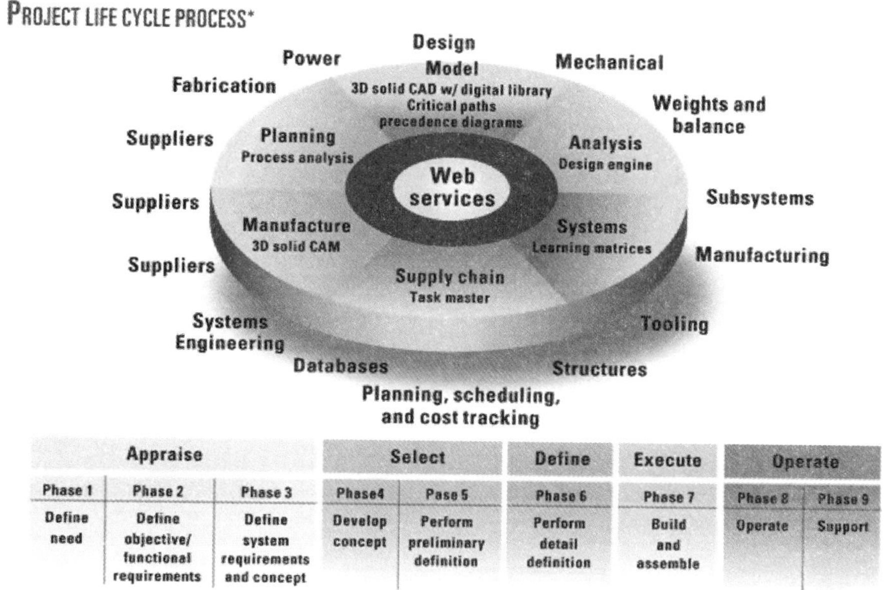

Fig. 4–3. The PLCM for manufacture of an airplane. Aerospace companies use Web services (hub at center) to connect planning, modeling, analysis, systems design, supply chain, and manufacturing as project proceeds (bottom) from appraisal (left) to selection, definition, execution, and, finally, operations and support for the vehicle for its life cycle (right).

Then, the world changed under the feet of both great manufacturers. After the September 11, 2001, attack on the World Trade Center, oil prices began a steady escalation, to hit $130 per barrel by mid-2008. Economy suddenly trumped size, the new prime metric became passenger miles per gallon of aviation fuel, and the 787 became the best-selling new commercial aircraft ever.

Boeing is a lean company. Boeing engineers swept through the PLCM, exercising real options for fuel economy embedded into the ISM for the 787. A traditional airframe was replaced by layers of carbon composites and aluminum foil. The wings became thinner, were swept upward, and were even lighter than ever. Suddenly, Boeing had the world's most economical long-haul aircraft, and they sold like hotcakes. In the meantime, Airbus struggled with its more traditional processes of manufacture. The world's largest airplane got stuck for two years on the cabin electrical system, of all things.

Modularity that Boeing built into the PLCM system, through its software-enforced distribution of the ISM, allowed it to manufacture the 787 all over the world and assemble it, like Lego blocks, in Seattle. However, the ISM missed the boat on two fronts: Boeing ran out of the tiny fasteners (the little screwlike plugs) that hold the modules together, and they misunderstood the complexity of plug-and-play assembly (the modules didn't quite fit together right); consequently, to their great embarrassment and cost, the delivery of the 787 was delayed. Boeing too is still pursuing perfection, but one certain fact is that their ISM will not forget the lessons learned from this first use of the modular techniques. The next plane design will not suffer the same problems.

In the oil industry equivalent, as the price of oil went from $15 to $130 per barrel between 1999 and 2008, the tremendous burst of oil field activity produced yet another oil boom. New construction of oil rigs suffered the same old problems of mismatched electrical schematics, routing and installation layouts, subsystems misfits, new-equipment downtime and other inefficiencies and shortages.

Version mismatches, incompatible modifications, and module conflicts could be identified in the model long before any construction or repair occurs if SE were prevalent throughout the industry. These tools and processes can be used to continuously appraise needs, reevaluate real options, execute the best design at the last moment, build, and then operate the asset for life (fig. 4–4). The oil industry has most of these tools and techniques, but not the systems integration capabilities that predominate in the aerospace and automotive businesses.

HOW LIFECYCLE MANAGEMENT INTEGRATES LEM ACROSS AN ENTERPRISE*

Lean energy management:
Transforms uncertainty into innovation to make systems better, faster, cheaper

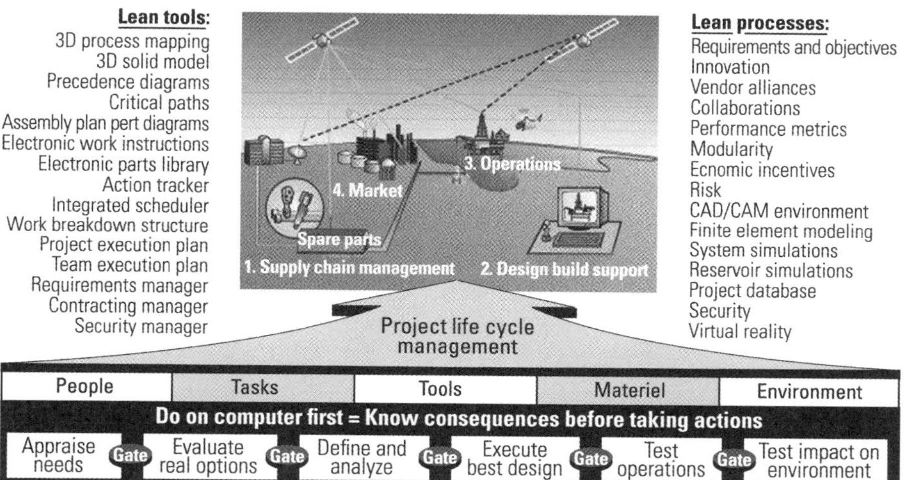

*Circles of Fig. 1 represent the enterprise.

Fig. 4–4. PLCM, integrating lean management tools across the greater enterprise. In the oil industry example (center), a long list of lean software tools (left) are used to make lean processes more efficient (right), so that people, tasks, tools, material, and the environment (bottom) are all integrated.

Engineering integration

The SE approach has proved to be far more efficient from both cost- and time-saving standpoints. It requires that all significant customer and contractor requirements be thoroughly understood before a specific solution is developed to address those requirements. All disciplines (structural, manufacturing, quality, security, support, etc.) participate in the SE analysis (fig. 4–5).

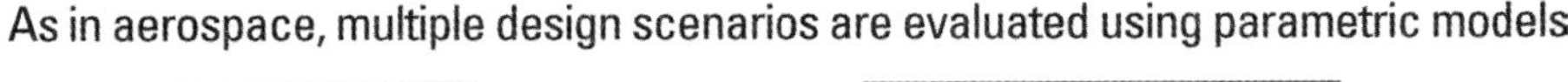

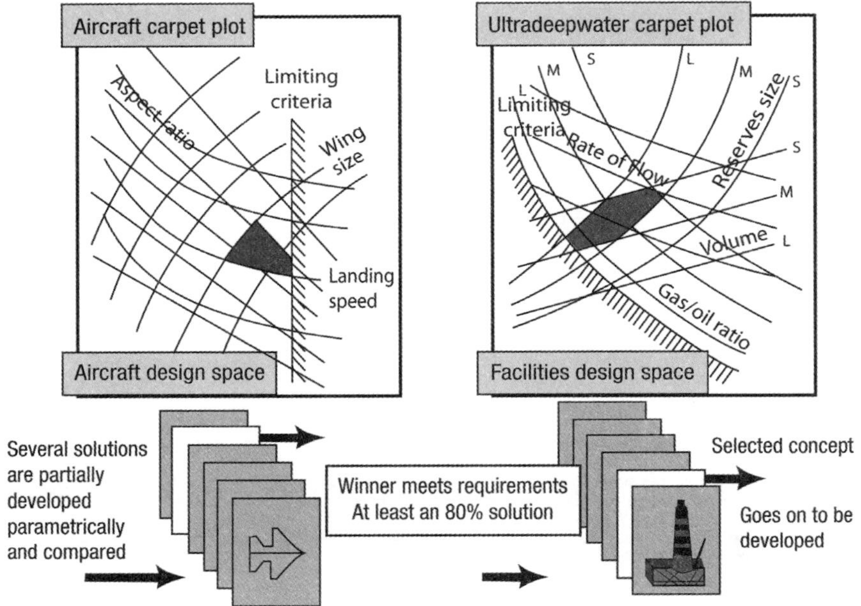

A "carpet plot" is used to define a "design window" that describes uncertainty of multiple critical variables. The deepwater "plant" contains uncertainties that must be accounted for that ultimately relate directly to the subsurface reservoir. These change dramatically over the life cycle of the facility.

Fig. 4–5. Illustration of carpet plot. SE requires that multiple scenarios are tested on the computer to determine the best design concept via multidimensional evaluations of parameters critical to performance. The optimal design space (gray at the center of both plots) provides the best value for the life of the asset, whether it is aerospace (left) or energy (right).

SE asks that multiple design scenarios be evaluated using parametric models—that is, parameterized families of probability distributions, one of which is presumed to describe the way a population is distributed. In PLCM, several designs are developed and scored for best performance. In aerospace, the limiting criteria that define the parameters are landing speed, wing size, and the aspect ratio of weight to size (i.e., how large it must be to carry a given weight). In offshore exploration, the limiting criteria are oil and gas reservoir size, the rate of flow allowed by the permeability of the subsurface rocks making up the reservoirs, and the volume of fluids that can be handled at the surface on the production platform.

A 100% solution is not required. Experience has determined that, in both the aerospace and the oil and gas industries, about 80% of optimal performance is the most economical from a cost and production standpoint. The tool that SE uses to determine this 80% solution in design is called a *carpet plot.* This is a plot of each of the critical variables for several design parameters. Figure 4–5 illustrates the carpet plot that determines what it takes to make small, medium, and large airplanes and oil fields. The space that satisfies 80% of desirability of design is also shown.

Modifications of one system—for example, reservoir size or landing speed—can be analyzed for adverse effects on other parameters. Conflicting issues consistently emerge among fabrication, weight minimization, cost, accessibility, maintenance, safety, and scheduling. As these conflicts are resolved on the computer, the form, cost, and performance of the likely optimal system design or modification slowly emerge. SE data flow starts with the customer needs, objectives, and requirements documentation. This leads to requirements analysis and carpet plots that evaluate the uncertainty window of the various competing designs.

A large-scale, full-system scenario is then simulated on the ISM to evaluate the responsiveness of the design to variances caused by uncertainties, such as price volatility or unusual component wear and tear. Real options are used to identify and assess the relative value of solutions addressing the uncertainties discovered by the simulations.

Feature-based design

The emergence of design conflicts leads to more in-depth computational analysis, or feature-based design, and integrated-systems analysis. These analyses focus on very specific areas, addressing unique issues that arise through simulations of variability. The critical conflicts usually cross disciplinary boundaries and involve three or more processes that in the old way would have been seldom examined jointly. Alternatives developed before the need for painful reworking is needed (fig. 4–6).

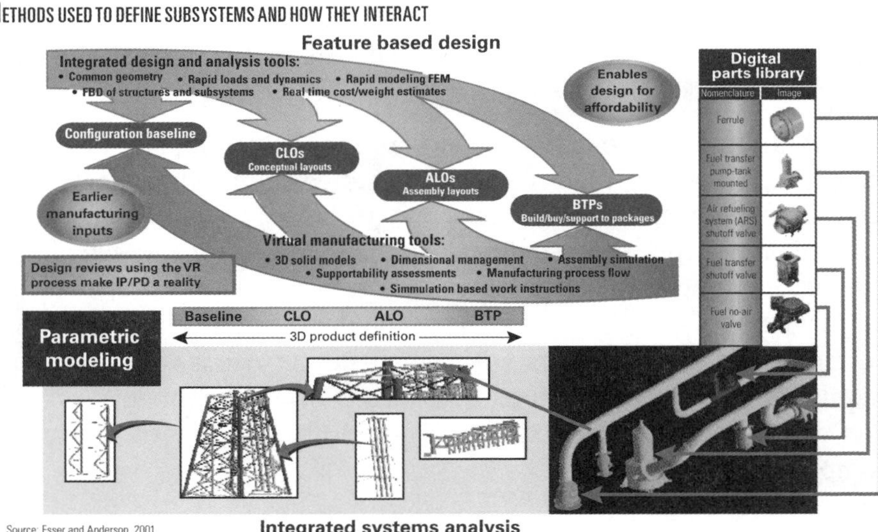

Source: Esser and Anderson, 2001.

Fig. 4–6. The feedback loop enforced by SE analyses

As the reconfiguration matures, many of the most difficult issues reappear. Electronic sign-off freezes features one after another until the preferred design emerges. A configuration baseline is defined in the feature-based design process, and then conceptual assembly layouts and build-to packages are designed using the digital parts library. These, in turn, automatically define supportability requirements. Integrated-systems analysis is used subsequently to test functional capabilities of the new assemblies, their ties to central control, and multiple-subsystem integration capabilities. Functional electrical schematics and logic diagrams are then automatically generated for each subsystem. A virtual prototype of the new configuration emerges, ready for the supportability simulation. The feedback loop enforced by SE (fig. 4–6) takes the design specifications and continually matches performance with manufacturing and operational realities, to develop an integrated-systems analysis of cause and effect for every component of the system.

Virtual supportability

Virtual supportability tools simulate the factory floor or field operations to lay out the most efficient construction and repair sequence for each project. Virtual supportability takes model scenarios and adds in operational and maintenance evaluations. It incorporates simulations of complicated maintenance tasks, including placing human-sized repair crews into a virtual-reality environment to ensure that tolerances will allow maintenance access (fig. 4–7).

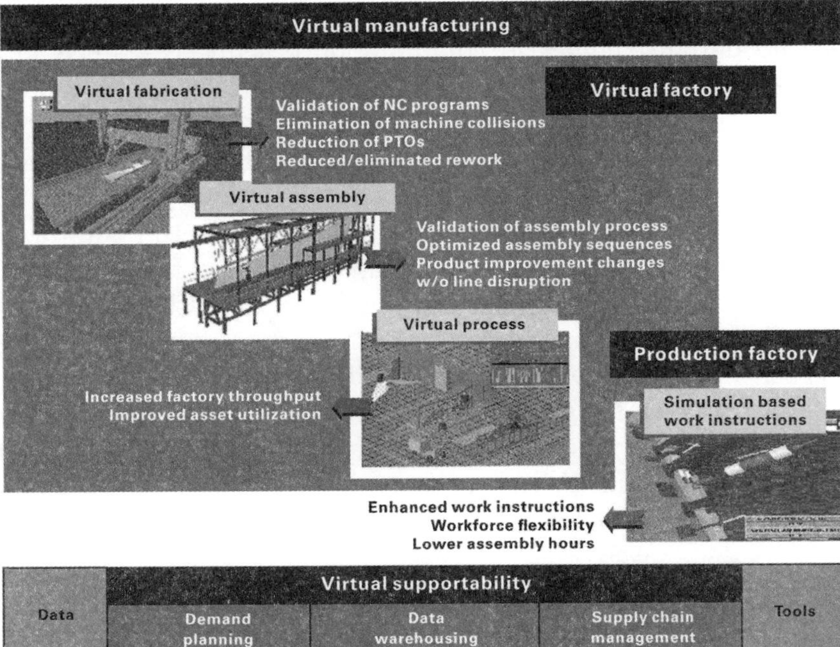

Fig. 4–7. Simulation of the fabrication, assembly, and operational processes, producing a supportability plan. Virtual reality is used to simulate the assembly line itself, not just the vehicle. NC: not connected (determinations of where design flaws do not quite fit); PTOs: production takeouts (components not absolutely needed).

Supportability plan

Simulation of the fabrication, assembly, and operational processes produces a supportability plan that extends over the life of the asset (fig. 4–7). The supply chain is laid out along with procurement sequences. Historical requirements, repair forecasts, and consolidation of inventories (turn rates) are used to determine appropriate sourcing and just-in-time delivery schedules. Virtual support begins the operator-training process, using simulated operational and maintenance tasks even as the construction is just beginning.

COST AND CYCLE-TIME GAINS

How can the energy industry effectively evaluate the risks associated with conversion to SE processes and tools? Until there is a substantial track record of profitability improvement, cost and cycle-time impact can be estimated by considering the following steps applied to all process subsystems:

- Use previous project experience of your personnel to describe the as-is state of project subsystem tasks.
- Define savings based on other lean industry performances at similar subsystem tasks.
- Add uncertainty factors for industry differences (these may initially be as high as 50%, but will drop over time).
- Add worst-case delays and confusion from first-time use of SE tools and processes.

We conducted a cost and cycle-time evaluation of the impact of SE on a generic, $500 million ultra-deepwater offshore oil and gas project. A suitability matrix was used to relate possible lean savings to the specific design/build segment of the project (fig. 4–8). In the early appraisal and selection stages, the aspects of SE that shorten the cycle time the most were

- Reduced design ambiguity.
- Faster review and decisions on design options.
- Posting detailed build instructions to all subcontractors far earlier into the schedule.

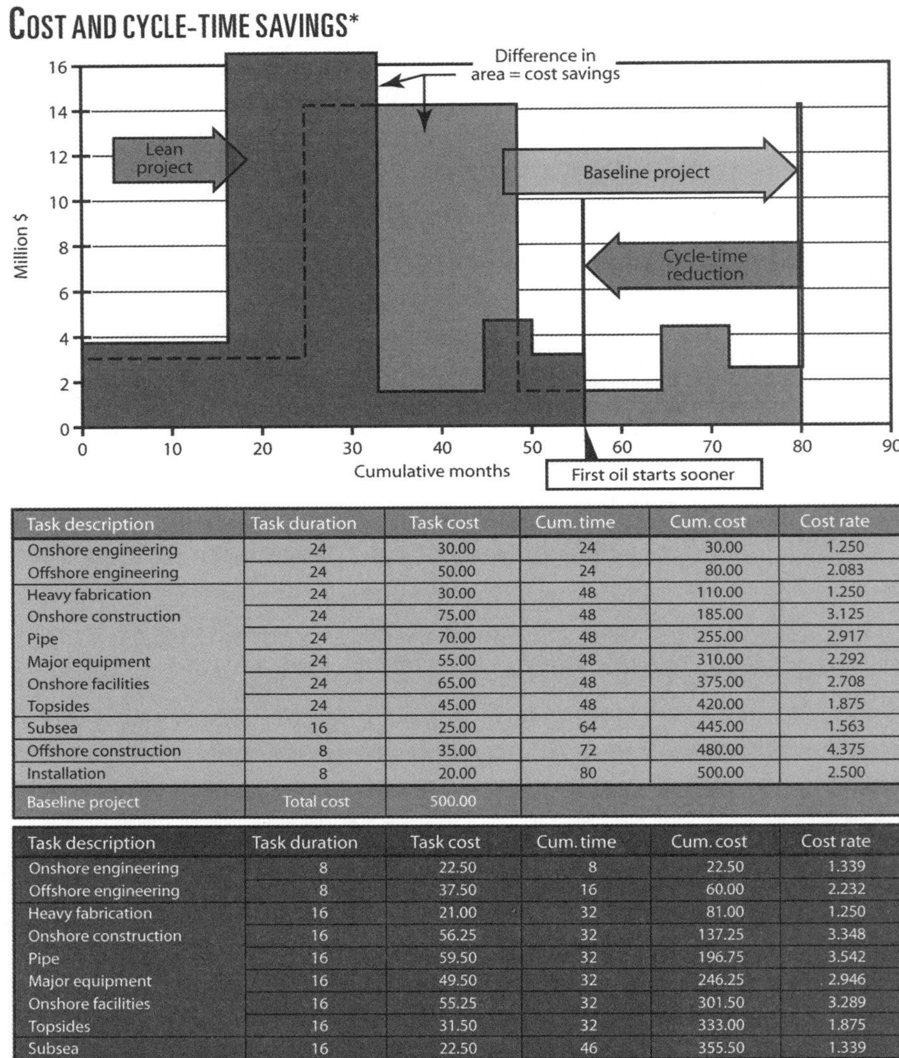

Task description	Task duration	Task cost	Cum. time	Cum. cost	Cost rate
Onshore engineering	24	30.00	24	30.00	1.250
Offshore engineering	24	50.00	24	80.00	2.083
Heavy fabrication	24	30.00	48	110.00	1.250
Onshore construction	24	75.00	48	185.00	3.125
Pipe	24	70.00	48	255.00	2.917
Major equipment	24	55.00	48	310.00	2.292
Onshore facilities	24	65.00	48	375.00	2.708
Topsides	24	45.00	48	420.00	1.875
Subsea	16	25.00	64	445.00	1.563
Offshore construction	8	35.00	72	480.00	4.375
Installation	8	20.00	80	500.00	2.500
Baseline project	Total cost	500.00			

Task description	Task duration	Task cost	Cum. time	Cum. cost	Cost rate
Onshore engineering	8	22.50	8	22.50	1.339
Offshore engineering	8	37.50	16	60.00	2.232
Heavy fabrication	16	21.00	32	81.00	1.250
Onshore construction	16	56.25	32	137.25	3.348
Pipe	16	59.50	32	196.75	3.542
Major equipment	16	49.50	32	246.25	2.946
Onshore facilities	16	55.25	32	301.50	3.289
Topsides	16	31.50	32	333.00	1.875
Subsea	16	22.50	46	355.50	1.339
Offshore construction	5	26.25	51	381.75	4.688
Installation	5	16.00	56	397.75	2.857
Cycle time and cost savings	Total cost	397.80			

*Estimated using lean suitability matrix methodology.

Fig. 4–8. Example of cost and cycle-time savings with a deepwater oil and gas platform project. Real savings from SE processes produce 50% savings in both cycle time and costs.

During the early stages of the project, digital design tools and methods had a much greater impact on configuration design than on supportability. Virtual manufacturing tools used for all stages reduced cycle time primarily in the execution stage. These tools improved the interfaces between subcontractors for hookup and commissioning, and in operations, through the shortening of the need for retrofitting. Supply-chain improvement was found throughout.

Cost impact was estimated by selecting value and savings on the basis of past project experience. These estimates were then refined through benchmarking metrics. This experiential base must be augmented by expertise from other lean industries—but with added uncertainty, because of possible industry differences (sometimes as high as 50%). Initial cost improvements were estimated through all stages.

For example, virtual support tools and methods provided significant operational cost reductions in both the definition and operation stages. In the definition stage, the most benefit came from improved tools and methods that deliver better visibility, clarity, and accuracy of interfaces to processes in which collaboration with subcontractors dominates. In the execution stage, virtual support tools provided better supplier and fabricator collaboration, improved visibility of the total product, and, in particular, improved change-order management, resulting in reduced errors.

We then calculated the cumulative cost and cycle-time savings over the life of the project, by summing cells of the lean suitability matrix (fig. 4–8). These were then compared to the baseline project to evaluate potential savings. In our study, the baseline project cost $500 million and took 80 months to complete. By our analysis, the same project done with SE tools and methods would have cost only $400 million and would have been completed in 56 months. In addition to these savings, the first oil would have been realized two full years earlier! The additional revenue from early oil alone would produce far more profit for the company than the $100 million cost savings.

COMPONENT MISMATCHES

Typical SE task flows for the design of an ultra-deepwater production platform project are shown in figure 4–9. The SE analysis starts with the gathering of customer needs, objectives, and requirements. In this case, the customer is the operator of the platform after it is constructed and installed. This leads to requirements analysis and industry-standards studies that equate to the geological work normally done by the oil company to find the oil and gas in the first place. The synthesis and control analysis is an ongoing feature of the process.

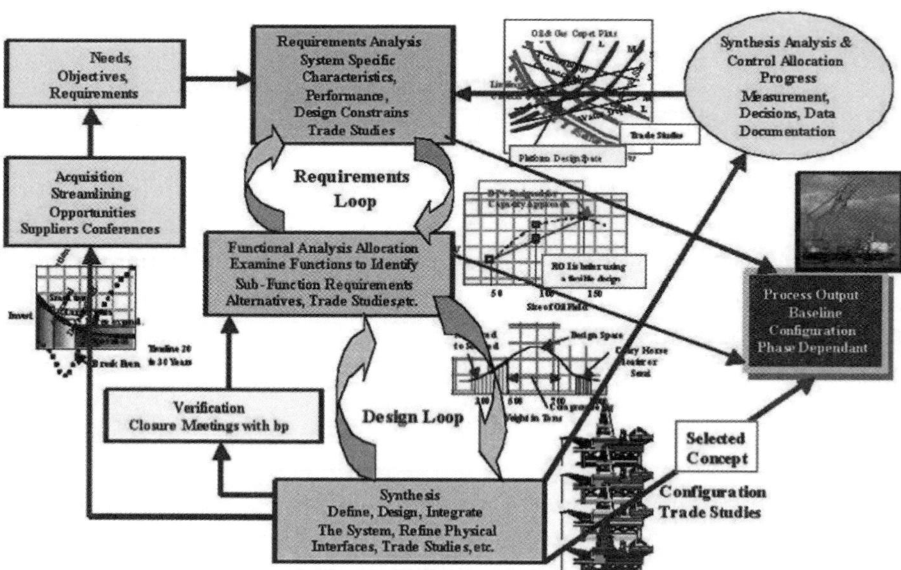

Fig. 4–9. An oil company's SE plan for construction of an ultra-deepwater production platform. Objectives and requirements (upper left) are iterated to produce an optimal design (upper right) that produces the best possible construction and operations. All parameters are optimized using the requirements and design feedback loops.

The requirements loop is next. Here, functional requirements go through a decomposition phase. The decomposition may touch on items such as full-field requirements (reservoir, processing, host facility, gas management, liquid export, etc.). These end in the design loop, where the actual concepts begin to take shape. A preferred concept emerges (wet vs. dry trees; floating production, storage, and off-loading ship; semisubmersible; subsea; spar; etc.).

The design loop then takes the working IPTs through a verification cycle, to ensure that the design addresses the initial requirements (low cost, minimal capital expenses, green solutions, earliest oil, and maximum recovery efficiency). Several hundred customer requirements are tracked and assigned to one or more cognizant disciplines.

Additional testing is then performed to verify that the design addresses the initial customer needs (strategic themes). Then, a supplier conference is held to identify possible bottlenecks in manufacture, and the design is modified. The next item in the flow-down is synthesis using scenario analysis. A large-scale scenario analysis might depict the possibility of finding more oil and gas than expected. Design sensitivities are then measured against a range of weather variations, distance from land, oil and gas composition, water invasion, method of transportation, and so forth. Other scenarios include early water breakthrough, unsustainable well production rates, and poor sand control.

Each segment is decomposed into functional threads that run through to operations, each of which is assigned to one or more teams identified using the requirements hierarchy chart (fig. 4–10). Procurement strategies and preliminary economic assessments are developed subsequently.

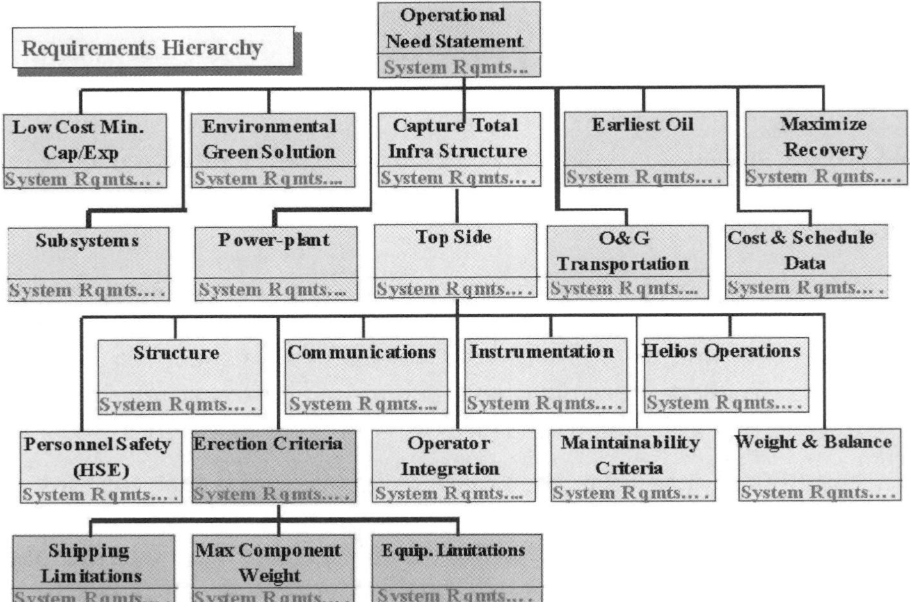

Fig. 4–10. Requirements hierarchy chart. The requirements hierarchy begins with an operational need statement, followed by costs, environmental footprint, and operational realities and ending with earliest oil (i.e., how quickly the first oil and gas can be delivered to market). This determines configuration and execution plans.

Design conflicts lead to more and more in-depth computational analysis. The critical conflicts are almost always interdisciplinary in nature, involving three or more silos of responsibility (fig. 4–11). The use of modern computational methods and advanced three-dimensional graphical interfaces focus the attention of the team on specific alternatives to be evaluated. The best or most balanced design option is selected and signed off on via a *design-decision memo* by all the project leaders. As these problems continue to be resolved, the design gains stability. As configuration maturity begins to crystallize, unresolved issues reappear but finally must be resolved, all before any physical work is begun.

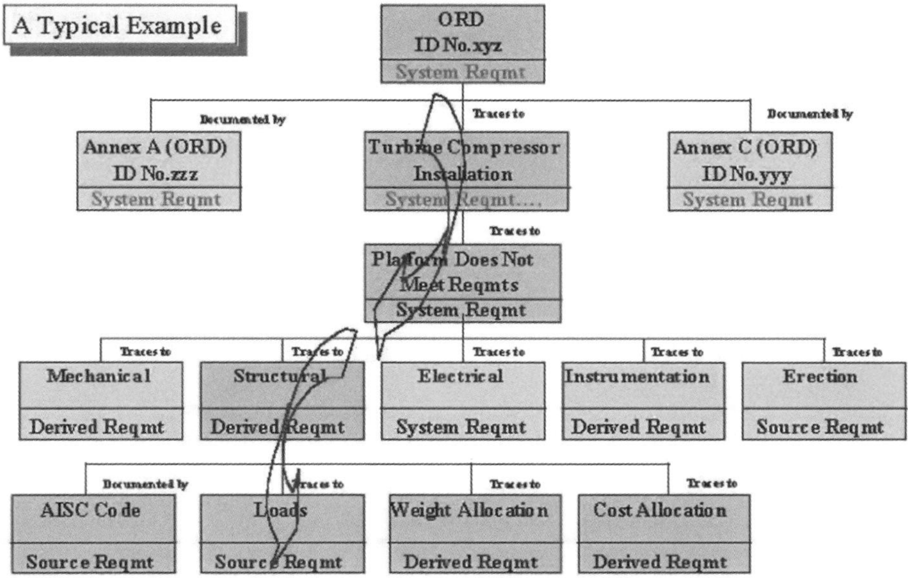

Fig. 4–11. Example of CALM techniques for tracking system requirements through all aspects of an offshore oil platform design. The ISM of the platform is used to identify the source of a structural anomaly. A compressor on an offshore platform is too big and located in the wrong place. This is discovered by tracking (the downward-plunging arrows) mismatches in loads, weight, and cost allocations—all within the model, before any physical work is begun.

A preferred design emerges, which is accompanied by extensive modeling and simulation, often in a modern, three-dimensional, virtual-reality environment such as a "cave." Daily or weekly IPT meetings of the affected disciplines are conducted to verify form, fit, and function. These meetings also serve to catch any design *dislocation* problems. These are captured using an action-tracker system that sends copies of stated problems to those involved in their resolution at the end of each virtual-reality session. At subsequent sessions, these problems are revisited. Each resolution is documented in a design-decision memo. The data and the tracking system are contained in a central database accessible to all, including subcontractors. Thus, decision-making does not always require meetings.

In summary, the economic value from the use of SE tools and techniques comes from

- Reduced instruction ambiguity
- Faster reviews of options
- Earlier scheduling decisions
- Better supplier and fabricator collaboration
- Improved visibility of the total project
- Improved change-order management
- Reduced site queries
- Better assessments of health, safety, and environmental impact
- Improved interfaces between contractors for hookup and commissioning
- Supply-chain improvement

The value of this transformation to SE should become evident when working through the preceding checklist. Most important, people must reinvent their workplace. All stakeholders must work together before the required cost and cycle-time savings from SE can be fully realized.

5 IMP/IMS

In this chapter, we suggest organizational, methodological, and management techniques that will help build an effective CALM implementation. These are neither rigid rules nor inflexible tools, but instead are intended to provide dynamic help as every project encounters surprises, difficulties, and other unknowns. The integrated master plan (IMP) hopes to identify the resources needed to accomplish a clear set of goals, and the integrated master schedule (IMS) tries to identify and keep track of the path for achieving them. They are designed to help with the understanding of the multitude of cross-cutting influences and desires that occurs during any project.

IMP

The IMP provides a suggestion of how to manage the integration within a project that is striving to produce a product or service, whether internal or external to the company. In short, the creation of an IMP is an attempt to better define what is to be produced and then to facilitate the communication and collaboration needed to get there. A summary of goals and objectives of the IMP is given in fig. 5–1.

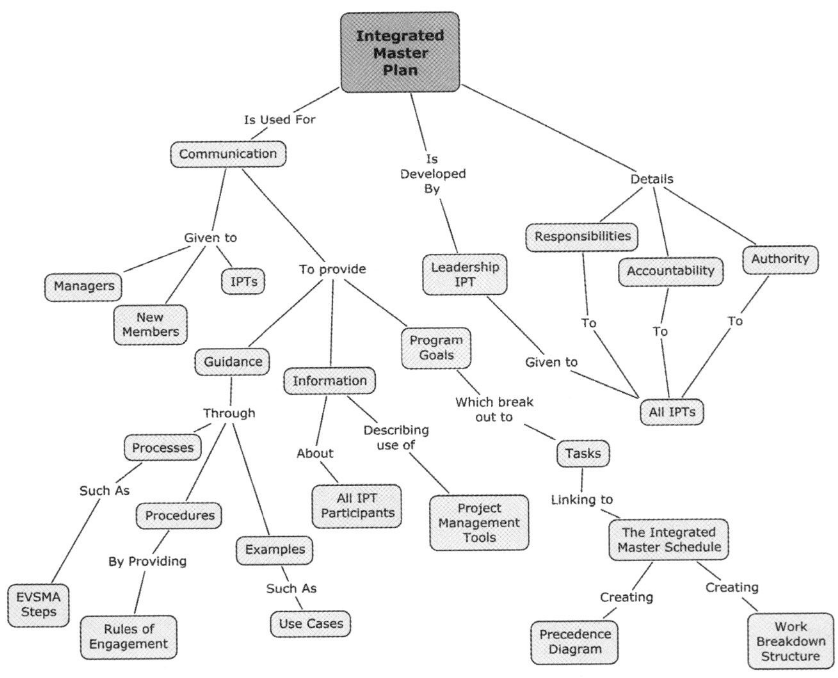

Fig. 5–1. Thought process provided by an IMP. This "mind map" tells a narrative, from top down, of how the IMP allows the problems to be organized and thought through to the end.

Methods

We recommend that the IMP be developed according to systems engineering procedures and methodologies. This entails identifying and developing strategies to arrive at solutions and requirements for executing the project and then generating the documents that record both successes and difficulties as the project proceeds. All levels of the organization need to be involved in the creation of the IMP, so that it becomes not a list of marching orders but a consensus of the ways to accomplish the goals of the project.

In reality, the IMP is not a rigid framework so much as it's about methodically planning and replanning the structures, systems, and practices of the company that need to be mustered to deliver the products or services of the project. It is intended to be a set of live documents that are updated as curveballs are thrown. An analogy could be made to the mid-20th century, when managers came up with printed forms to fill out for project management. Back then, carbon copies of these forms were used for the purposes of coordinating the progress of each department toward the execution of the project. Nowadays, the documents of the IMP are usually spreadsheets that are electronically published as they are developed or updated.

Integrated process teams

The IMP is executed by cross-functional working groups called *integrated process teams* (IPTs). These full-time IPTs are made up of users, business analysts, IT experts, and engineers from many different departments to execute the major segments of the project. The IPTs empower individuals, so that they have responsibility, authority, and accountability (RAA) to carry out the subprocess or asset that they are in charge of deploying. IPT leaders should know systems engineering, so that they can understand and foster the efficient running of each IPT and manage the uncertainties inherent in all projects.

There are two levels of IPT (more detail is provided in fig. 5–1):

- *A leadership IPT.* Traditionally, this comprises the sponsors and project management, with functions of managing the IMS and the IMP. All of the leaders of working IPTs are included in the leadership IPT. The leadership IPT makes modifications to the IMP and the IMS; provides staffing and funding for the working IPTs; resolves conflicts; and conducts weekly briefings, monthly reviews of progress, and annual, 5-year, and 10-year planning sessions.
- *The working IPTs.* Examples of primary working IPTs are work process, operational management, and ISM teams. There are also specialty support IPTs that answer to specific technical needs. These support IPTs come and go as needs emerge and are satisfied.

Each working IPT is charged with definition of its own requirements, communications with the customer, affordability, and integration with others during the planning stages of the project. When the execution of the plan begins, they are responsible for definition of the economic value metrics to be used, integrated-schedule coordination, risk management, and other metrics and management processes that evolve as the project proceeds (fig. 5–2).

We recommend that each IPT begin by defining the plan (fig. 5–2, left). Customer communications come first and foremost—that is, listen to what the customer wants and needs, rather than plan based on what the IPT thinks they need. Next, define the requirements and affordability of the project. Reach agreement with the customer on how to measure program progress and success at the IPT level, define perceptions and expectations, and review requirements. Establish a measure for customer satisfaction and pay attention to it. Share the vision, mission, and goals of each IPT plan.

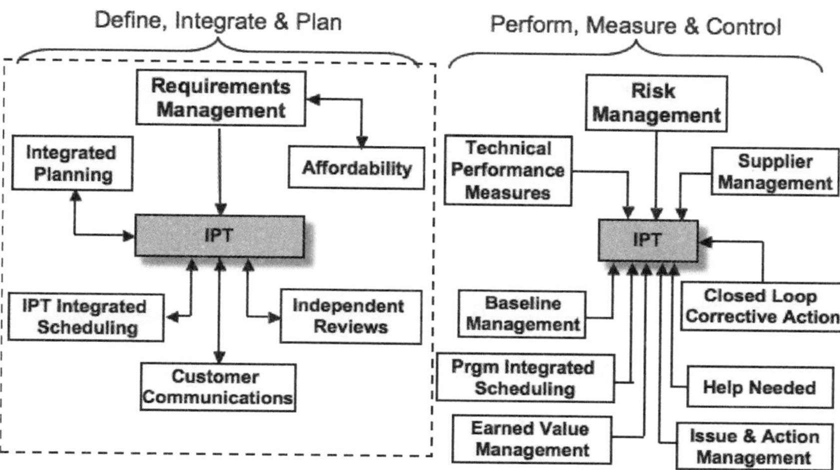

Fig. 5–2. Metrics and management processes for which IPTs are responsible. IPTs are charged with RAA for defining, integrating and planning each project, and then executing the planned steps, measuring the outcome, and in many cases, controlling the operations after the project has been built.

Next, the IPTs move to the execution of the plan (fig. 5–2, right). Set up the risk management infrastructure first. Communicate the good news, as well as the bad. Establish technical performance metrics for yourselves and your suppliers. Deliver a baseline to management, so that they can track your progress against it. Establish feedback loops. Automatically update the master schedule daily and weekly. Develop earned-value management criteria and manage actions. Be responsive to customer reactions. Treat all people with fairness, trust, and respect, and most important, work together rather than hold too many meetings. IPTs coordinate the development of requirements, systems, and changes to processes that are in the IMP by way of status reviews (fig. 5–3).

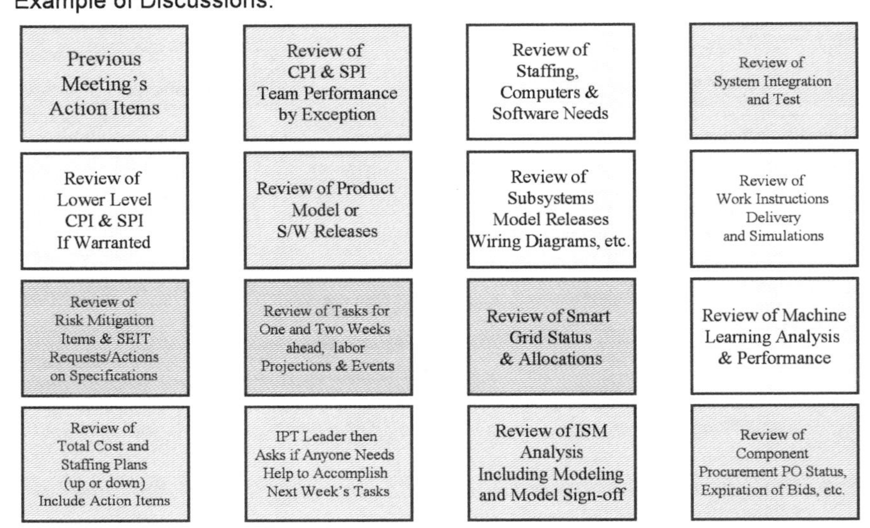

Fig. 5–3. Status reviews. IPTs review weekly the who, what, where, when, and why of the project, so that efficient project management and execution is maintained.

System Development Process

The IMP is at its heart a system development process plan. This concept combines the best practices of systems engineering, requirements engineering, and lean software development. In the engineering literature, there are many views of the system development process. Some focus on the many stages of requirements analysis and definition, while others focus on the stages of the product life cycle. Other views emphasize the modeling required at each level to validate the requirements and the development of independent testing programs that will be used to verify system performance and acceptance. Our goal is to suggest that each company take the views that best meet the challenges presented by the realities of your own projects. But in general, the system development process flows from the top-down with requirements definition from management (fig. 5–4, left) leading to functional, physical, and design definitions by the IPTs (fig. 5–4, right).

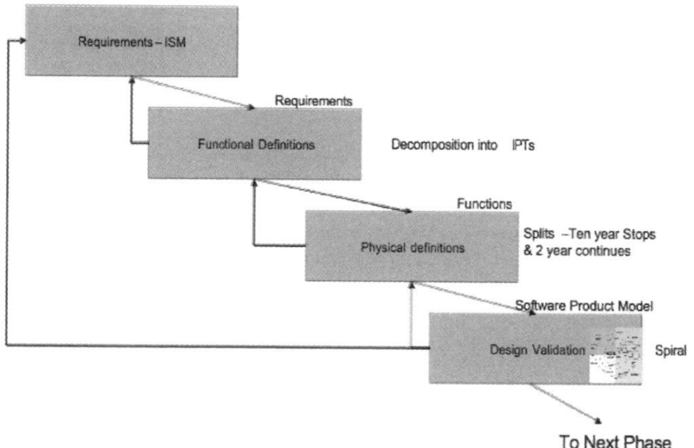

Fig. 5–4. System development diagram. Top-level goals from management (left) are converted by the IPTs into the steps in the iterative process to develop the system base on well-understood engineering principles (right). The spiral at the bottom is discussed later (see "Requirements Definition"). (***Source:*** E. Hull, K. Jackson, and J. Dick, ***Requirements Engineering,*** 2nd ed., London: Springer, 2005)

Once an IMP has been created, some initial requirements can be stated to the IPTs by management. The IPTs then further define the requirements and map the system development process using BPM to capture the customer's short-term as-is condition and long-term to-be objectives. These statements of customer needs are then translated into system requirements. At this point, the set of requirements are defined that will guide the IPTs toward projects that are feasible to attempt in the near term and will allow them to plan what is needed for long-term tests of new innovations (fig. 5–5).

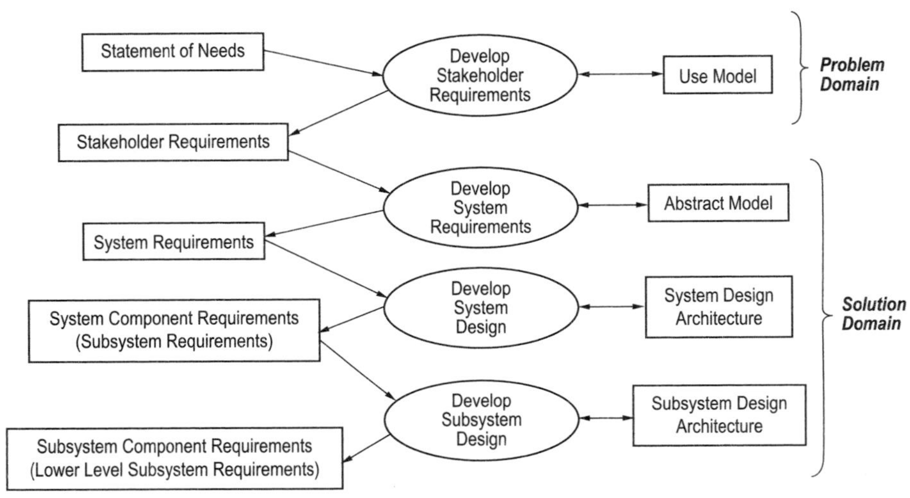

Fig. 5–5. System development process. Once projects are identified and approved by management, they proceed to the requirements definition process, where the statement of needs will have been developed through extensive interaction between the IPTs and the customer. (Source: Hull et al., ***Requirements Engineering,*** 2005)

In an ideal world, the IPTs will begin with the development of the system requirements without constraints imposed by other existing subsystems and processes, to understand what the true needs of the customer are. An as-is map of the system will also be developed. An assessment will be performed to determine the replacements and improvements needed to produce a to-be map. Replacements and improvements will be assessed against the objectives of the overall project requirements—namely, reducing costs and cycle time, as well as increasing reliability. As these near-term projects mature, the IPTs continually re-evaluate them to ensure that they are not only meeting the near-term requirements but are also addressing the long-term requirements. Modifications to the top-level goals often require significant changes in the long-term requirements, as well as assessment of how well these short-term projects support the evolving long-term goals (fig. 5–6).

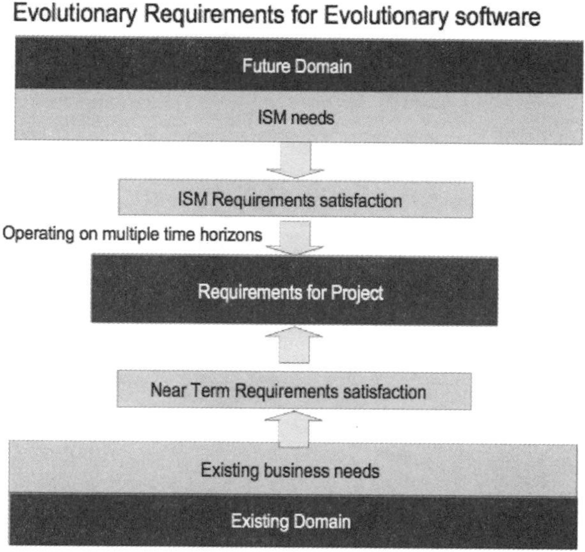

Fig. 5–6. Requirements development for the ISM. The future (above) and the present (below) merge at the center to form the requirements of the project.

The to-be domain represents the state where we would like to end up in future years and how that state can be used to derive requirements for improvement down to the system level. During the design process, modeling and analyses are performed to guide the design process by using the formal requirements as inputs to this process. For example, the ISM can be used to model, at various levels of the requirements hierarchy, goals/usage (including BPM), functionality, and performance (fig. 5–7).

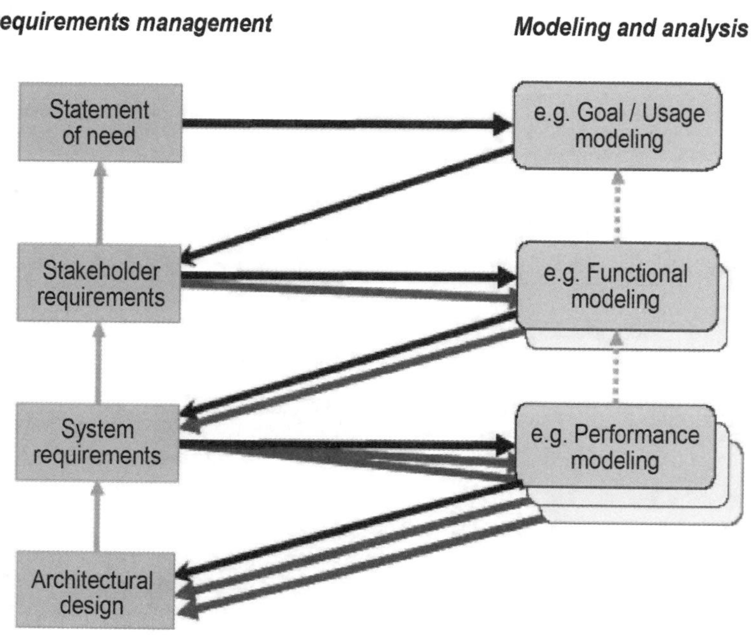

Fig. 5–7. The management of requirements form the basis for determining whether the system to be developed will actually meet the requirements specified earlier in the design process (***Source:*** Hull et al., ***Requirements Engineering,*** 2005)

The requirements form the basis for building test suites that are used to validate the behavior and performance of the system. Testing of prototypes at all levels against the requirements helps to ensure the quality of delivered systems, in terms of both what they do and how well they do it (fig. 5–8).

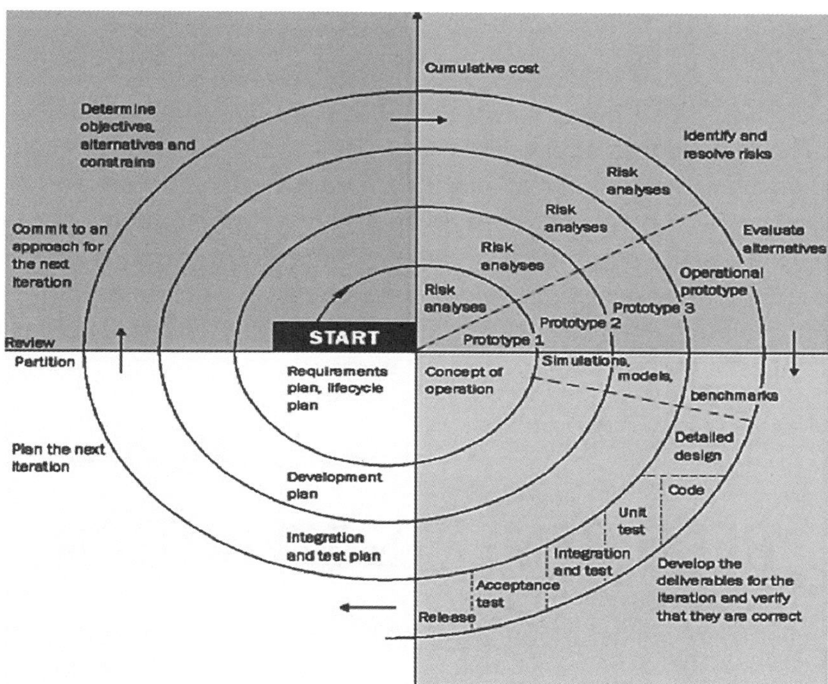

Fig. 5–8. The evolution of a CALM project. Repeating loops for prototyping migrate the project to operational readiness. (***Source:*** Hull et al., ***Requirements Engineering,*** 2005)

The systems engineering approach of using both the near-term and the long-term requirements works because most products such as software can be structured to be an iterative development over time—as long as optionality is preserved in the beginning—to meet the longer-term goals. A more expedient, lean type of requirements mapping, such as agile software development, can be used for the near-term solutions, to address existing business needs, while a very robust systems engineering approach can be used for the long-term requirements of the overall goal. The reality is that you never reach all long-term requirements; rather, you expand the realm of opportunity and continuously evolve the IMP to keep focusing on what needs to be developed in each near-term cycle, as it rolls by, to enable the future development of the ever-evolving business. Thus, the IMP becomes a live business plan.

The IMP should always have a long-term view, thus also enabling R&D to be focused now at places where new techniques and tools will be needed in future years. In this way, CALM implements long-term goals while continuously tracking short-term payback and ROI. Strategies change over time, and so will the requirements as the future vision evolves,. Thus the IMP minimizes the risk of not meeting the challenges of the future (fig. 5–8).

REQUIREMENTS DEFINITION

Requirements definition is the first step of the system development process. It answers the following questions: Why are we doing this project? What business/customer needs does the project intend to satisfy?

Statement of needs

In any project, the need for a new product or service is expressed in terms of saving money, reducing risks, improving productivity, and increasing customer satisfaction (reliability). Goals are set by management, but they are not generally the intended users of the product or service. However, the management constraints placed on the systems—particularly those currently being used to run the business—are an important contributor to the requirements that will be developed. In other words, the statement of needs includes the need to continue to run the business successfully. As stated elsewhere, the energy business is not building an airplane on the ground, but is instead trying to improve the one we have as it is flying.

Stakeholder requirements

During the process of developing stakeholder requirements, it is important to develop a set of scenarios and use cases that help in the development process, by describing the user's needs in language that is easily understood (i.e., in the language of the problem domain). Use cases are scenarios that examine in more detail how a user might actually be using a system. They also identify screens, forms, and controls needed to effect the decision-making that new software would address. The value of these scenarios lies in the clues that they provide to yet-unstated requirements. These scenarios do not represent the requirements themselves, because they are intended to illustrate only a set of particular paths through what may be a large set of conditional decision points. They are invaluable for identifying the points at which decisions are being made. Also, they can guide the system engineers in the design of tests that can be used to exercise the system.

From this analysis, we decide on the projects that will be undertaken to meet these needs. Consider the need for new maintenance software as an example of a functional area, to illustrate the next steps in requirements definition. From maintenance operations, we hear that they need tools to help them perform more effective preventive maintenance to improve reliability. They also need to produce better schedules for work crews, to reduce wasted time and thereby address the need to lower costs. Use cases and scenarios begin the development process.

Functional definition

At this level of requirements definition, the functional areas that need to be modified or enhanced are mapped. Functional analyses and allocations are the processes where the basic building blocks are specified. Each of the system functions that must be addressed is identified, and the requirements of each function are established and recorded in the requirements-tracking system. Once these system building blocks are identified, it becomes possible to perform trade-off analyses to assess which of the many alternatives best satisfies the requirements against measures of the costs and complexities introduced for each alternative. At this stage, the integration of these building blocks is undertaken to identify the interfaces between functions that are required and the external interactions with other systems or people that will be required in the near and long terms.

System requirements

At this level, we transition from the problem domain to the solution domain. We develop requirements for possible tools or business process changes that could address the stakeholders' needs. Continuing with the maintenance example, consider the types of tools that might lead to improved reliability. These include analytical tools to help identify components that would benefit the system most if maintained better, and database tools that, coupled with advanced sensors, better monitor and report the health of those components. Computer-based schedulers might then be able to remove wasted time from the system and establish better data communications between supervisors and technicians, to reduce errors and inefficiencies.

Physical definition

Specific hardware and software are then identified that can provide the concrete solutions to deliver the required functionality. As before, alternative approaches are explored and evaluated in this phase. Trade-off studies to evaluate the effectiveness of particular solutions should be done. Note that this is the point at which we evaluate total life-cycle costs. What costs would a solution incur from the perspective of initial purchase, customization, integration, and maintenance? Other dimensions besides cost—such as future flexibility, scalability, and risk—should also be considered. We should ensure that

- All viable alternatives are considered.
- Evaluation criteria are established.
- Criteria are prioritized and quantified when practicable.
- Subsystem/components requirements are not fully defined.

At this level, the system is further broken down into the types of subsystems or components that would need to be assembled to solve the system-level problem. For an analytical tool to assist in preventive maintenance, databases of historical data and current operational data for components are needed. An ML algorithm that processes the data and identifies and prioritizes maintenance activities is then tried out using the historical data for testing at this stage. Reports or operator interface tools are designed to communicate analytical results, so that actions can be taken. Finally, a data collection system that measures the results of the actions taken, to determine the effectiveness of the new system, is necessary. At this level, particular scheduling approaches and vendor packages are also evaluated and chosen.

R&D requirements

Questions are then asked about what new R&D is required for the long-term success of the project. In our maintenance example, perhaps new dynamic scheduling methodologies or ML techniques are required. Use cases and scientific proposals are good ways to flush out the requirements, but it is essential to put the customer directly in contact with the researchers so that the R&D will be properly focused on solving the required problems.

Design validation

For large, complex systems, it may be necessary to build a system test environment that can be used at many stages of the design and development process, to perform unit and system testing before and after the system reaches the production environment. In many critical applications, as much additional money may be needed for designing and building the test environment as the actual system. Even when applications are deemed less critical, it is mandatory for the requirements developers to define the tests that need to be developed at each level of requirements definition, to ensure system compliance.

Acceptance plan

The acceptance plan will be developed after usage studies are complete. Within the collaboration framework, an acceptance plan will be part of the requirements development. This plan indicates what is an acceptable solution that satisfies the user's problem as discovered in the problem-domain analysis. Normally, the acceptance plan is detailed between the development of user requirements and the development of functional requirements.

VALUE ANALYSIS

One of the critical components of the CALM methodology is the measurement of performance indicators as a means to enable continuous improvement in operations. Many popular themes exist in the business world to formally state how value will be achieved through changes in the way we plan and do work. For example, consider Hammer's reengineering ideas; Deming's plan, do, check, act (PDCA) cycle; and Six Sigma's define, measure, analyze, improve, and control (DMAIC). These all have in common an adaptable organization that can sense a need to change and then make the change before it is too late. Our ability to sense a need to change requires the ability to cheaply and efficiently measure both performance and the business environment. An adaptable organization measures the effects of the changes, maintaining those that succeed and abandoning those that don't perform well.

At the initiation of any project that attempts to change its environment, there must be a way to sense whether results have been beneficial or detrimental, as well as whether the change was implemented in the first place. Without implementation of both the change and positive results, success of the project cannot be guaranteed. A good example of the key performance categories for CALM processes is shown in figure 5–9.

Each IPT will be responsible for the development of performance indices that will provide the baseline metrics to define success from specific initiative completions. Part of the leadership IPT's responsibility will be to review the creation of these metrics and to monitor the success of these metrics. Any proposed initiative should be measured in terms of the value that can potentially be achieved and the metrics that will be used to monitor the success of attempting to achieve this value.

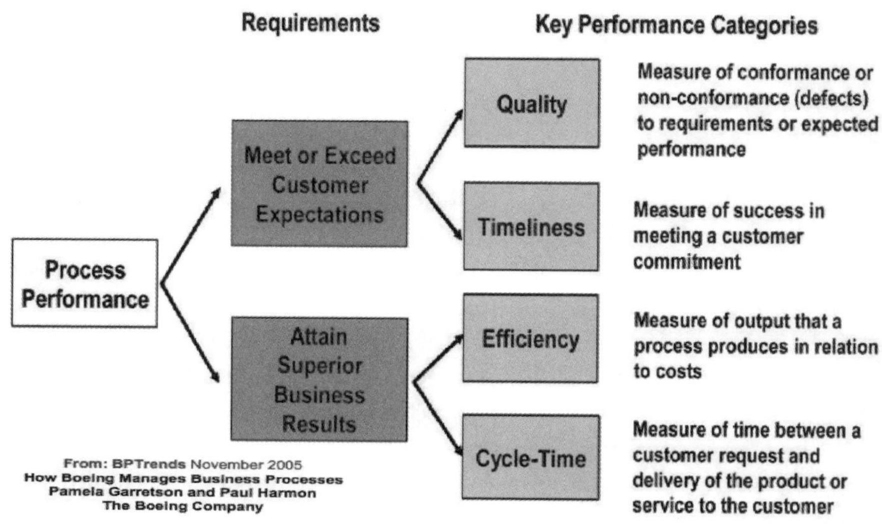

Fig. 5–9. Performance tracked all along the way. Questions of whether customer expectations have been met are answered using quality and timeliness metrics. (***Source:*** P. Garretson and P. Harmon, How Boeing A&T manages business processes, ***BPTrends,*** November 2005)

There are different types of indicators. One relates to how savings, such as costs per widget or time spent per widget, can be minimized, while another is more directional in nature and tends to confirm levels of success. Other indicators include volatility mitigation and measurements that document better-managed business risk.

There will always be more great ideas than resources available to chase them down. (Figure 5–1 contains a template for submittal of new projects to the leadership IPT.) Thus, value analysis is critical to assignment associated costs and performance metrics to track progress prior to any decision to move forward (fig. 5–10). This, in turn, helps prioritize which projects will support the optimization of resources and result in the biggest bang for the buck. Enterprise value stream mapping and analysis (EVSMA) provides an example of one useful template to prioritize and execute future work.

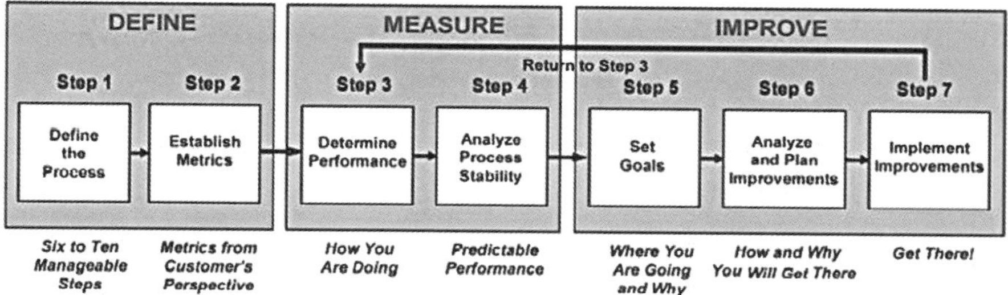

Fig. 5–10. Progression in process improvement. Steps migrate from left to right. Once step 7 has been achieved, improvement continually loops back through steps 3–7.

IMS

We recommend that an IMS be created that combines the plans of all the IPTs. It should contain a timeline of all work that is being planned within the IMP. Updates on work progress should be made regularly. A *precedence diagram* (fig. 5–11) is a very effective tool for keeping track of critical sequences of work, such as what needs to be done before what can be started. Metrics are then developed to flag potential issues that arise with the schedule and the associated resources don't match.

Microsoft Project, Primavera, or some other common management tool can be used to keep an electronic, modifiable, continually updated, and version-controlled schedule of the various sequencing required in order to keep the project on schedule. The IMS should have resource loading, cost monitoring, and risk level tracking.

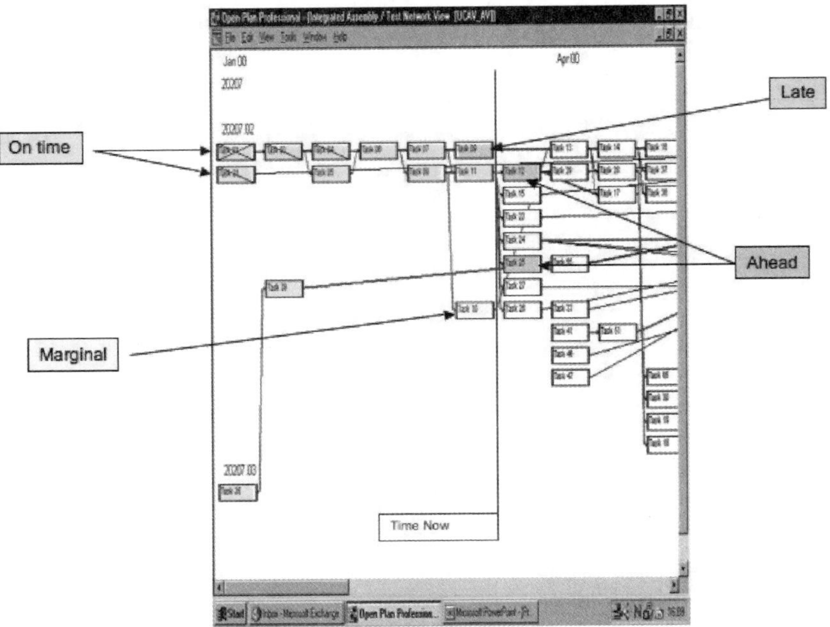

Fig. 5–11. Exemplary precedence diagram. In the IMS, functional groups provide the metrics, IPT leaders distribute hours on the basis of metrics, and team performance is reviewed weekly against the precedence diagram.

Each IPT should have its own schedule, deliverables, milestones and goals, and *critical paths* (CPs). The precedence diagram keeps track of the sequencing required. The precedence diagram creates workflow paths through the IMS and then tracks progress and identifies when and where deviations from the CPs are occurring. Management can respond before train wrecks occur in the urgent delivery of value to the company.

The precedence diagram uses the CP to construct a model of the IMS, including the following:

- A list of all activities necessary to achieve the goals
- The time (start date and duration) that each activity will take to completion
- The dependencies between the activities
- The cost of each in manpower and dollars

The IMS maps the starting and ending times of each activity between goals and tasks, determines which activities are critical to the creation of value, and defines the activities critical to developing the product or service. The CP then separates those activities or tasks with critical delivery times from those with *float times*—activities or tasks whose timing is less critical. Any delay of an activity on one of the CPs has a direct impact on the planned project completion date (i.e., there are no float times on the CP).

Since the project schedule changes on a regular basis, the CP allows continuous monitoring, allows the project manager to track the critical activities, and ensures that noncritical activities do not interfere with critical ones. In addition, the method can easily incorporate the concepts of stochastic prediction, using program evaluation and review technique (PERT) charts.

Gantt charts provide a graphical display of project plans, and can be used to track planned and actual accomplishments. Microsoft Project and Primavera allow one to show interdependence, resource allocation, and roll-ups of tasks and subtasks into summaries on a Gantt chart, with varying degrees of clarity.

In summary, the IMP provides a virtual model for a project so that task managers can refine their plans as realities change the plan. The IMS provides a timeline of the plan and project status to the team and the customer. The IMP and IMS entail greater effort than the standard Gantt chart, but they are more helpful to the effective execution of a project in that they assist in determining creative reactions to problems as they occur. The methodical approach that is contained in the IMP and the IMS is similar to that required for NASA to plan, build, and execute the first moon landing (see Box 5–1).

BOX 5–1. ORGANIZATIONAL NOTES

IMP goals

- Get everyone on the same page.
- Communicate rules of engagement among teams.
- Provide processes and procedures to effectively accomplish project goals.
- Write down the project goals and associated tasks.
- Understand individual tasks as they relate to the Big Picture.
- Enable coherent understanding of logistics/game plan.
- Identify who does what, where, when, and why.
- Explain how to work together via integrated process teams (IPTs).
- Provide documentation requirements.
- Identify the management tools that are to be used and specify why.
- Risk mitigation plans are developed.
- Responsibility, authority, and accountability (RAA) are identified throughout the project.

IMP objectives

- Interfaces and RAA between and among leadership and the working IPTs are established and signed off on by all.
- All major activities are clearly aligned with project drivers: health, safety, and the environment; cost and value measures on all levels; schedule and time as value measures; and quality control.
- The IMP is dynamic, constantly being revised and updated according to progress and problems encountered by the various IPTs.
- Plans are integrated for engineering, process change, enforcement, metrics (what does and does not work), procurement (field and control-centers), plan modifications, and start-up.
- Key project team leaders are identified who are responsible for the execution of the detailed plan.
- An IMS is created for all IPTs that reflects the timing and coordination of execution of the IMP.
- Organizational charts and responsibilities for all levels of the project are created.
- Cost performance indices (CPIs) and schedule performance indices (SPIs) are measured and managed, to ensure that the program is properly resourced to accomplish all tasks on time and budget.
- A framework is established to continually bring problems to the fore as system concepts and designs evolve.
- Impacts and risks are minimized through systems engineering integration and testing (SEIT).
- A process is created to facilitate decision-making for schedule changes, issues tracking, version control, and conflict resolution.

IPT responsibilities

- Leadership IPT:
  - Mission: The leadership IPT is made up of top management and the leads of the IPTs, whose mission is to provide a channel for communication and coordination across the various IPTs. It also serves as a window into the functioning of the IPTs for the company's executive management team. The leadership IPT acts as a clearinghouse for acquiring the resources that the CALM project needs to execute the overall mission.
  - Scope: The leadership IPT has visibility into all the IPTs. The leadership IPT has weekly conferences to go over the prior week's progress and issues. Each IPT is responsible for development of an IPT-specific outline summary, to be submitted to the leadership IPT prior to each conference.
  - Interfaces: Each IPT will have representation on the leadership IPT to communicate its progress and issues.
  - Measures of success: Milestones in the ISM are tracked to verify that they are being met on a timely basis and within budget. Customers are engaged to determine if they are satisfied with progress toward their perceived value creation from each CALM project.
  - Organization: This team is made up of at least one representative from each IPT, plus a select number of advisors, including R&D and IT.
  - RAA: The team is responsible for ensuring that the activities of the all the CALM teams are coordinated with each other and with the relevant management personnel. They are also charged with delivery of information to executives.

- Working IPTs:
  - Mission: Based on the statement of work (SOW), the working IPT will write down what you intend to do. For example, "This XYZ widget IPT is responsible for the design, analysis, procurement, fabrication, testing, and qualification of the following products. This IPT will support system integration at the installation and investigate A, B, and C. This IPT will interface with O, P, and Q IPTs and provide the following database to the IMS and the IMP." Development of the XYZ widget itself is not only documented, but the work is also then executed.
  - RAA: What are you responsible for? An example would be the development and production of a new type of widget for some XYZ system. What is the extent of your authority? For example, does the IPT leader have the authority, as delegated by the program management, to obtain the necessary resources (people, materials, tools, etc.) and to accomplish the tasks described in an approved SOW. What are you accountable for? An example would be the on-time/on-cost delivery of some agreed to number of widgets to the XYZ customer on some specified date, at or below a budget of $RR million.
  - Milestones: Include a chart of the major milestones that the IPT will use in planning their individual activities, along with the major milestones that the IPT expects to meet during the first year.
  - Risk: Each IPT will establish a process for mitigating medium to high risks. It includes identifying the likelihood and consequence at each level for suppliers and all teams and functions. Risk mitigation plans

are drafted, analyzed, tracked, reviewed, and mitigated at regular intervals until they are low risk. Risk management eliminates uncertainty, lessens consequences, and allows the team to prepare and implement contingency plans. It minimizes or eliminates many risks. It highlights areas of uncertainty and false confidence. It helps in deciding the best course of action.

- Technical performance measures (TPM): Each IPT will establish TPM. Key product performance capabilities are identified. Key lower-level performance measures are identified and allocated to program IPTs. The program then calculates, allocates, and tracks technical parameters given to subordinate teams. Measuring technical performance shows progress relative to satisfaction of customer requirements. TPM pinpoints emerging lean achievements and deficiencies.
- Affordability: Each IPT will establish an ongoing assessment of a program to ensure that it is being executed within customer planning and funding guidelines. It must have sufficient resources identified and approved and is managed on the basis of accurate cost and staffing data. Affordability requires an established program affordability plan and RAA, with targets for cost and effectiveness, to achieve customer satisfaction.

IPT documentation

- Value analysis and baseline metrics—to ensure prioritization and focus of effort.
- Business process model (BPM)—to model the way you work.
- Case studies—to improve the user comprehension, which is required in order to make correct decisions.

- Functional requirements—to communicate and agree on what is to be built.
- System design diagram—to communicate and agree on what performance is needed from the system.
- Test and acceptance plan—to commit to what is expected.

All documents are kept in a digital data repository and available for all to see at all times (including subcontractors).

New project submissions to the leadership IPT

- State the problem:
  - Create the problem statement for a new project.
  - State the goals of this project and explain how these goals help achieve the CALM goals in the IMP.
  - State the potential benefits that the execution of this project should bring to the enterprise. Be specific in terms of costs in time and dollars.
  - State the champion (i.e., the person with the RAA) for the project. Identify the resources needed to accomplish the project during the following phases: discovery, analysis, prototyping, and implementation.
  - State the estimated time frames for each phase.
  - Develop the use case studies for this project.
  - Create a mission statement.
  - List the goals and subgoals that are needed to achieve the mission.
  - List the stakeholders (i.e., participants) who are going to achieve the goals.

- Develop scenarios describing how the participants will operate the system to achieve these goals.
- Develop an analysis plan that will be used to verify that achieving the project goals will indeed assist in achieving the enterprise goals.
- Describe stakeholders, processes, and metrics.
- State the scope of the project, as well as what is beyond the scope of this project.
- Identify the stakeholders associated with this project and what impact this project has on the stakeholders.
- Map out the as-is work processes, in BPM, that have an impact on this project's goals.
- Identify the data, as defined in the analysis plan, that will be used to assess the performance of the current system/operation.
- Identify the metrics that will be used to determine whether the project's goals have been achieved when the project is complete.

- Construct current-state perspectives:
    - Collect artifacts that can be used to define the current state of affairs in the system being examined. For example these can include procedural manuals, training documents, and, in many cases, forms and screen shots that capture current user interactions with the system.
    - Develop scenarios describing how the system is currently used and how it interacts with the rest of the enterprise.

- Identify enterprise opportunities:
  - Assess the opportunities for moving the system toward the project goals by developing a model of what the current state is and what the future state might be, based on the collected data and documented processes.
  - Assess whether the goals of the CALM project are being achieved. By comparison between the current-state and the future-state models. Reveal differences where waste and inefficiencies in the current state can be addressed to increase enterprise value.
- Describe the future-state vision:
  - Develop a vision statement describing how the future state will look and behave.
- Create a transformation plan:
  - Analyze the differences between the current state and the future state, to develop the requirements for the future-state system.
  - Enumerate, for example, any new systems, new data collection, and process changes that will be required.
  - Describe the products and services that will be delivered by implementing this project.
  - Develop a plan to produce and integrate these new capabilities into the existing organization.
  - Estimate the costs to implement this project.
  - Develop a refined benefits analysis for implementing the project.
  - Define a set of metrics for each step in the transformation that can be used to ensure that the goals of the project are being met.

6

BIG PICTURE

A CALM system continually seeks perfection in performance through enforcement and monitoring by real-time software systems. Aggressive learning for such continuous improvement (termed *Kaizen* by Toyota) requires feedback, which is a key concept of CALM. The goal is to reach Kaizen through rigorous enforcement of feedback loops that first predict outcomes and then make corrections based on objective scoring of the blind tests of success of the predictions as actual events unfold. To create the business capability for continuous improvement, we are striving to migrate from information to knowledge management through modern process control.

The evolution in knowledge management can be illustrated in terms of games. The TV game show *Jeopardy* can be won consistently if information is managed efficiently. If it weren't for the time constraint from the buzzer, a computer could win every game. The only requirements are a good electronic encyclopedia in its information database and a Google-like search capability.

Backgammon is a step up from *Jeopardy* in complexity. While a computer can still be trained to win every game, it must be programmed to deal with the uncertainties introduced by the throw of the dice. Moves must be recomputed on the basis of a changing board position. These games illustrate the progression from information management (*Jeopardy*) to knowledge management (backgammon).

KNOWLEDGE MANAGEMENT

The energy industry can and should adapt its decision support systems to the new knowledge management paradigm. Progress in three dimensions is required. In the future, we hope to simultaneously convert information into knowledge—while providing decision support through modeling and simulation—and close the feedback loop by adding adaptive learning to our enterprise management system (fig. 6–1).

KNOWLEDGE MANAGEMENT IN 3D

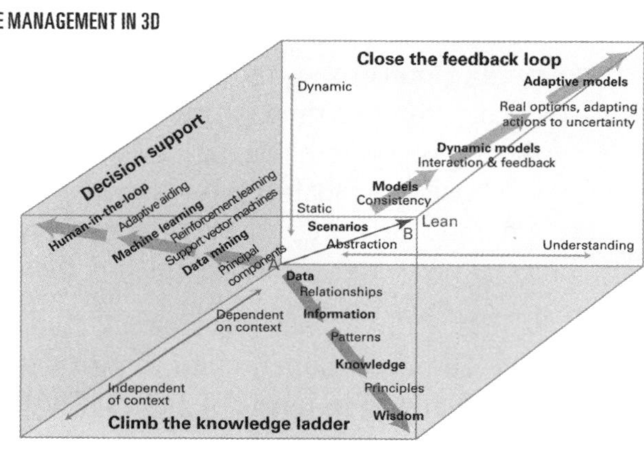

Fig. 6–1. Knowledge management. The three-dimensional trajectory from point A to point B along the diagonal of the knowledge cube requires skills and software for simultaneous management of information, decision support, and the feedback loop. (***Source:*** Modified from a two-dimensional figure in M. Whiteside, The key to enhanced shareholder value, Society of Petroleum Engineers paper 71423, 2001)

The knowledge cube

Modern process control is a multi-dimensional affair, which we present as a *knowledge cube* (fig. 6–1). Every business is always dealing with its position within the cube that simultaneously projects to points on all three backplanes at once. Consider the ideal progression of the intelligence of a business. A lean business will progress from the lower back to the upper front of the cube, from a static view of the world dominated by attempting to use data to determine context to a dynamic, adaptive organization whose employees act with wisdom and experience (represented by migration from A to B in fig. 6–1).

On the contextual lower surface of the knowledge cube, the company will be migrating from data management to information management, allowing relationships among data to be defined. On the next step up this ladder, patterns are recognized that convert the data into more and more actionable information in regard to system functions or correlations. Nevertheless, patterns are not knowledge. Knowledge requires the involvement of people. Many businesses have failed to recognize that humans are required in order to transform information into knowledge. While there is still a role for knowledge management tools and best practices captured at this level, a core concept of the lean approach is that knowledge lies fundamentally in the wisdom of people.

Therefore, progression from point A to point B in figure 6–1 also engages decision support to keep humans in the loop. For attainment of the conversion of data into wisdom, data mining progresses to ML and adaptive aiding so that better and better decision support occurs in the business (left backplane of the cube in fig. 6–1).

At the same time, computer models must evolve from just providing possible operational scenarios to being capable of modeling whole-system performance (rear backplane of the cube in fig. 6–1). Consistency and linkages among the many silos of management responsibility need to be modeled first, before they can be understood and implemented. Then, models need to migrate toward being dynamic, rather than static. The same models that are developed for operational calculations (e.g., power flow, reservoir simulation, batch mixing, and business processes) are used for dynamic simulations of future outcomes.

A business can thus prevent catastrophic incidents from happening—anticipating them through model simulations. These models can be used to fully understand the dynamics of a business through simulation. In addition, simulations can be used for training of personnel in how the system reacts and how they, in turn, should react in the face of a future dynamic event. Improvements in efficiency can come from exploring the better operations simulated on a computer via thousands of scenarios to optimize the business on the computer. As uncertainties change, the dynamic models are able to evolve with respect to the changing events and become adaptive to the environment.

Perhaps the least understood concept of the knowledge cube is how these three planes interact within the space of the cube. The natural progression is from understanding data and information to modeling the enterprise and adding new ML technologies that learn from successes and failures, so that continuous improvement not only is possible but is the operational dictum. The energy industry has enough data and sufficient modeling capabilities to evolve to this new lean and efficient frontier.

Finally, progress up the knowledge cube requires that this process of data analysis, modeling, and performance evaluation be done continuously. Then, the system will be empowered to improve itself to create value to the enterprise and prevent potentially catastrophic events from happening to the system. Altogether, the knowledge cube enables a company to see the competitive landscape, predict the unknown, and wisely run the business.

Digital convergence

Processes in the energy business are complex, not only because of the physical size of the supply chains required to get the products and services to market but also because of the complex array of uncertainties that can affect decision-making along the way. For example, in the electric power industry, operating engineers and managers are faced with a complex array of variables arising from quasi-competitive electricity markets, uncertain weather conditions, variable customer demand growth, fuel supply costs, equipment failures, environmental regulation, and the threat of terrorist attacks.

Energy industry uncertainties have impact on business processes that cross management silos and scale across many orders of magnitude of cause and effect (e.g., local to global, speed-of-light to weekly, and surface to miles beneath the surface). Further, the impact of each of these variables is dependent on the state of many other associated variables. Energy systems exhibit nonlinear behavior that must be managed by a correspondingly sophisticated, computer system that is focused simultaneously on operations, process, risk, and the creation of value to customers and the enterprise. Otherwise, organizations risk simplification to justify their decisions.

Conventional industry process controllers have dealt with complexity by fragmenting decision-making, by partitioning it into organizational divisions and hierarchal levels (e.g., separate regulatory, supervisory, and planning silos). For example, tactical control of the manufacturing process might be centered in one office while strategic investments, operations, and capacity planning are carried out in another office. The latter decisions affect business processes on longer timescales.

Business decisions made at lower levels are often made in isolation and on the basis of static assumptions that are out of date with conditions at higher levels of the organization. This fragmented approach leads to gaps and missed synergies that affect the efficiency and security of the business process. In such a broken process, opportunities to improve enterprise value through achievement of strategic goals and objectives are lost.

A real-time control system to prevent this backsliding from happening is now possible owing to the digital convergence of several simultaneous and disruptive computer and communications revolutions: wireless, cheap plug-and-play sensors (IEEE 1451), Web services, cheap distributed processing and storage, and hyper-exponential scaling of the capacity and affordability of memory and processors that exceed Moore's law. (Moore's law describes an important computer hardware pattern that has held fast since 1970: the number of transistors that can be inexpensively placed on an integrated circuit is increasing exponentially, doubling approximately every two years.)

The digital convergence of these new technologies enables industry not only to drive down the costs of traditional SCADA systems by an order of magnitude but also, in the foreseeable future, to convert to an automated, real-time two-way control system. We anticipate that the digital convergence will enable *innervation* (comparable to the human nervous system). The control system of the future might use Web services as its most critical distributed component, with a grid computing system layered on top to host the real options and RL feedback loops that will allow innervation of the field assets. For example, storage network technology can be used to distribute the traditionally centralized, IT department–controlled, data historian functionality of SCADA to the critical field devices themselves by use of time-series data-tag storage. In other words, each asset will carry all the relevant historical data about itself in its own enhanced radio frequency identification (RFID)–like chip in the field. Transformers and compressors might come with their own sophisticated chips as the modern car does today—in fact, this technology is now emerging in the energy industry.

Adaptive learning can then be built into these chips so that they become ubiquitous, distributed controllers of themselves. In other words, learning is designed into the system at the most fundamental, "last-mile" level, instead of just in the control-centers. Simultaneously, metrics are kept of all actions of the system so that continuous improvement becomes the norm.

Innervation is a key to creating distributed infrastructure security as well, enabling consumer marketers like Wal-Mart to require RFID tags on all products on their shelves. The devices, embedded with the intelligent controllers by the manufacturers directly into the products, will have the ability to broadcast their location and state of health from transport to sale and beyond. Theft will become more difficult, and more resources will be required in order to penetrate and affect such a distributed system.

Most of the components in this type of digital equipment will soon be commercially available to an energy company—see, for example, the Watchdog-Agent concept of the National Science Foundation Industry/University Cooperative Research Center on Intelligent Maintenance Systems. The proper development of requirements will allow staged application of this technology to specific uses, with a "crawl-walk-run" implementation strategy in mind. The methods of CALM will point an energy company toward coupling of these new sensor technologies with lean control software, to dramatically improve efficiencies and reduce business risk through two-way, sense-and-respond interactions among and between field assets.

Such a continuous, 24/7 CALM nervous system will likely incorporate adaptive algorithms and systems engineering processes at their most fundamental level and track the health of each component over time. Each field asset—such as a natural gas compressor, or a transformer—will then "understand" (via cheap, embedded silicon) its impact on business choices through fine-grained real options, using extensions (e.g., IEEE 1451 electronic specification sheets) that include the business side of decisions that can be ubiquitously made by each asset. The cheap silicon associated with the widgets will allow for the distribution of intelligence of the digital nervous system outward from control-centers to all field devices (fig. 6–2).

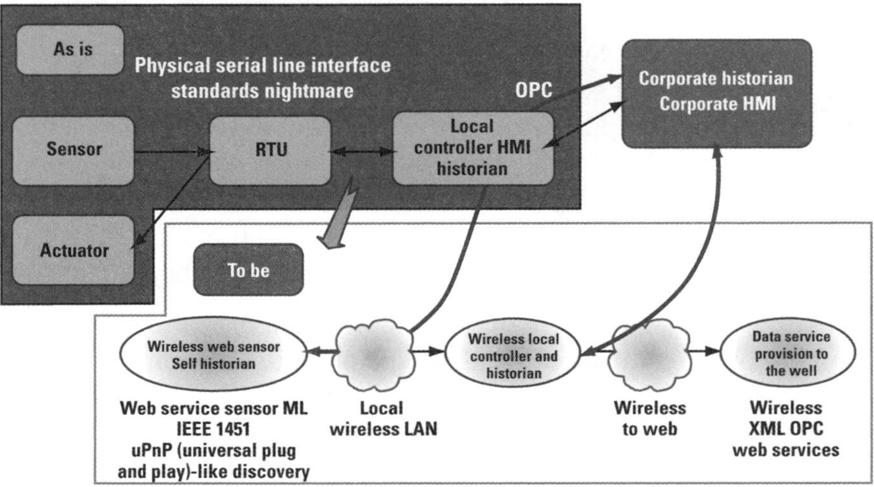

Fig. 6–2. The as-is condition of SCADA networks (upper left) and the to-be transformation with wireless clouds covering major cities (lower right). RTU: remote transmission unit; OPC: Object linking and embedding (OLE) for Process Control; LAN: local area network; HMI: human/machine interface.

RL controllers will eventually become a fundamental property of the distributed nervous system infrastructure (fig. 6–3). RL controllers will provide optimal decision-making as uncertainties over time are sensed; real responses of the system are differenced from expected results and are fed back to drive continuous improvement. They will function as the brains for the distributed nervous system. In other words, learning will be designed into the system at the most fundamental level.

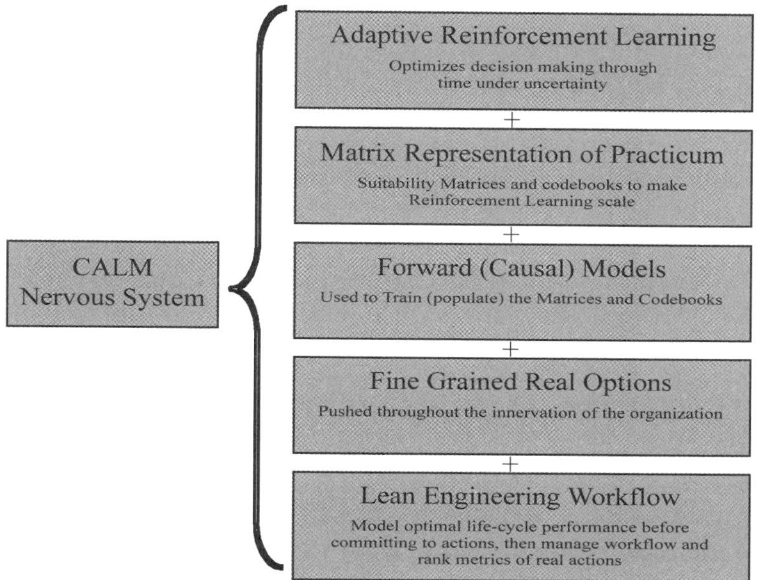

Fig. 6–3. Ingredients enabling a CALM RL controller to simultaneously manage RL, suitability matrix mapping, model simulations, real options, and workflow management of field operations.

In the future, adaptive RL will operate at the highest level of the stack of technologies that combine to produce the intelligent RL controller, which in turn will decide the actions and policies of distributed systems like oil and gas fields, pipeline networks, and the electric grid (fig. 6–3, top right).

Practical experiences (*practicums*) can be captured and loaded into a set of matrices that link problem columns with rows that contain symptoms such as variances from expected behavior of the subsystem.

The PM provides the simulation capability for forward modeling to test for root causes (fig. 6–3, center right). Scenarios represented by many different simulations will be scored, and those that successfully produce the most improvement will be used to control the weights that populate the cells of the matrices. The simulation results can be used with ML function approximation to help scale RL. They can be archived, so that results can be called on quickly by the RL controller to interpret and help determine actions to correct problems identified by the analysis of variances from field monitoring.

Real options for evaluating the financial value of the actions that the RL controller is considering will be computed at the site of each problem by the equipment involved (fig. 6–3, center bottom). We call it *fine-grained* because the real options are pushed out into the field assets themselves, so that they know and continually reevaluate their own worth to the system.

Last but certainly not least, there must be lean workflow in the field to execute the actions of the RL controller (fig. 6–3, lower right). The major addition to existing workflow management systems (WMS) of most companies is that the results of the work are scored for how effective the actions were at solving the problems that they were meant to address.

Configuration of the RL controller

To optimally manage the production and delivery of any energy product into and through the supply system to customers, the distributed RL controller of the future will have to be able to scale sufficiently to provide a unified decision support system that balances the realities of business constraints and contractual requirements with the uncertainties of the field (weather, price, supply, demand, equipment failures, etc.). Such a vision can currently operate on a small scale (e.g., one pump jack or compressor station or possibly load transfer in a small electric network) to optimally manage the portfolio in real time, but a great deal of R&D is required before enterprise scale can be attempted.

The CALM business focus in the meantime will be to begin with processes that connect business decisions to the last mile of the supply chain (the customers themselves) on a small scale, such as providing distributed intelligence for electric meter reading. Even this small-scale focus requires that several challenges to process connectivity must be overcome (the smart-meter example will be expanded upon to describe the complexity involved):

- The information architecture to integrate and optimize real-time business processes will need to scale from the very smallest of components (each electric meter) to the largest of markets (the hourly fluctuations in the price of electricity on the open market).
- The information trail needs to be ubiquitous, extending from headquarters to every field agent and asset (connecting the meter reader, call-center, accounting, and billing).
- The decision support system should be designed to maximize service but minimize cost to the customer, while simultaneously evaluating risks and benefits to the health of the electric grid, all in real time (load relief decisions at each meter have to balance customer comfort and safety with grid stability and resiliency).
- The underlying business logic should be transparent, so that it can be easily communicated both to customers and to new generations of operators and managers.

PUTTING IT ALL TOGETHER

The RL controllers will always seek to eliminate the wish-I-could-have-seen-it-coming events by computation of predictive failure models, cost-benefit analyses, and risk-return optimizations to identify the most efficient remediation solutions to problems quickly. All subsequent actions need to be tracked to verify not only that the work was done on schedule and on cost but also that the outcome was beneficial, as predicted.

Ultimately, the simultaneous achievement of corporate business goals and customer satisfaction through intelligent planning and execution is a feedback control problem. In the future, CALM will bring time-critical processes online in real time so that the feedback control problem can be solved properly. This will involve real-time data feeds from the field, spot and futures commodity pricing, data historians and data warehouses, and software applications that have not even been conceived in today's world.

We are beginning to grasp the scale and impact that such intelligent, distributed systems can have on the future of energy businesses of all kinds, as well as their significance to their ultimate customers. The future management of an intelligent, distributed system is a *credit assignment* problem of equating the actions of individuals (customers and employees alike) and subprocesses (involved in the creation and delivery of products and services) to overall corporate and societal goals. The credit assignment problem is solved using innervated software tools like RL controllers and hierarchies of them (fig. 6–4).

THE BIG PICTURE

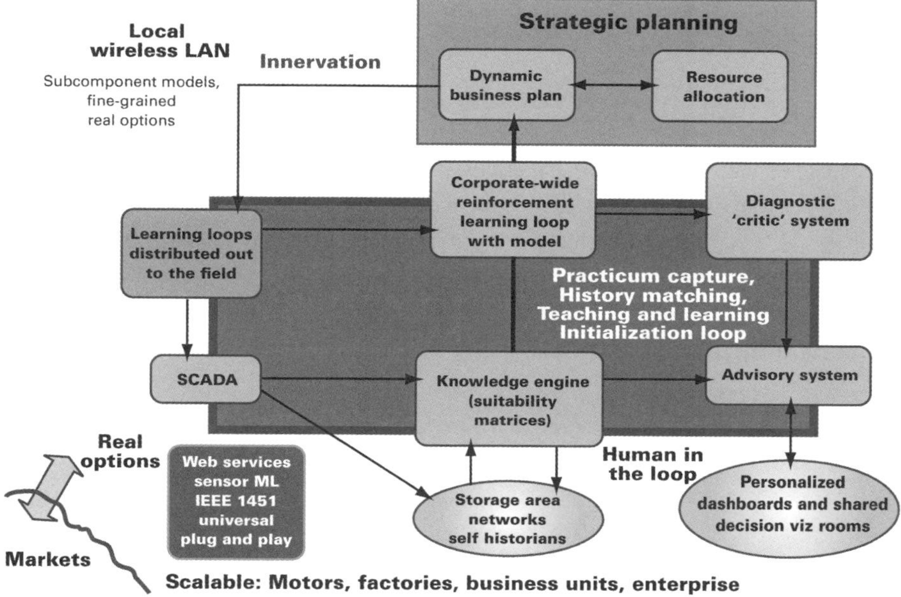

Fig. 6–4. Embedded and distributed computational infrastructure of the Big Picture. This is the formative vision of a peer-to-peer field network that is aware, in real time, of changing uncertainties, so that it can continuously choose new policies and actions that support profit-at-all-times while always giving the customer satisfaction while heeding operating, safety, and environmental constraints of the enterprise.

There also must be an enterprise-wide activity model used to represent all the business process activities in sufficient detail to materially contribute to weights in credit assignment, information dissemination, information management, task management, best-practice capture, and data-mining functions. Only then can dynamic schedules be fully optimized, resource constraints of people and equipment choreographed, and supply balanced with demand—all while business risks are evaluated and responded to continuously. In the Big Picture, RL controllers will have to generate actions that are optimized with respect to financial profitability, engineering efficiency, and customer satisfaction at all levels of the business process (fig. 6–4). In particular, the central "Corporate-wide reinforcement learning loop with model" in figure 6–4 is not achievable with today's technologies even with the digital convergence. CALM efforts, however, are working toward this goal.

Even though this Big Picture is futuristic, reinforcement learning and appropriate dynamic programming are being successfully applied to problems at many organizational levels in industry today. These applications already include regulatory control, supervisory control, scheduling, operational planning, capacity planning, portfolio management, and capital budgeting with real options.

7

ADDITIONAL TOOLS

In the preceding chapters, we saw that achieving implementation of CALM is a never-ending quest to improve performance. In box 7–1 at the end of this chapter, we list a wide variety of additional CALM tools that are available to help this journey along. All are intent on improving efficiency in some way. Mapping which of these your company is already employing is a starting point for determining the path to the lean transformation of your company. Next we describe additional tools that we have found valuable to create the business capabilities to recognize what improvements to do, and in what order.

SUITABILITY MATRIX

We use a knowledge elicitation method called the *suitability matrix* to build the lean practicum for populating the simulator that is used to develop the best scenarios for creating the to-be state of the company and its processes. The suitability matrix captures and encodes knowledge from best practices elicited from company experts and knowledge bases. The suitability matrix toolset employs 3D visualization and Web-based graphical user interfaces to efficiently collect feedback from the experts in the company.

We introduce this tool by providing a simple operational example from the oil and gas industry. We then show the progression, for a given contingency, from identification of symptoms, to the connection of those symptoms to problems, and then from the problems to solutions. Composition and hierarchy are derived through the matrix formalization itself, and that can eventually be used to guide a RL controller for optimal decision support.

Most companies have technologies that issue alarms when out-of-variance performance is identified. In alarm systems, variances are usually identified by *manage-by-exception* rules and by data mining. Alarms can be assembled from any of the following: physical device failures reported over SCADA systems, maintenance warnings, planning bottlenecks, electronic work instructions based on a workflow system, variances from plan, past performance history, best practices, historical analogs, and key performance indicators (tracked as each job proceeds, as well as afterward). What's missing is understanding if there is connectivity among the problems coming in from the field. With CALM, understanding interconnectivity starts with the suitability matrix.

Matrices are powerful tools for evaluating the connectivity between variables by using familiar form of rows and columns. Matrices are indeed simple to use, and by chaining them together, you can build rule-based expert systems. You can evolve these into more advanced ML tools, which in turn allow more integrated connectivity and quantitative rigor. In the meantime, matrices provide very powerful tools for representing, mapping, and modeling lean solutions.

For example, a matrix can be used to represent a generalized functional network, such as the evaluation of an oil and gas prospect. In the simple suitability matrix shown in figure 7–1, an oil company is trying to decide where to develop their next oil and gas field from among many opportunities that have been discovered and wildcatted (i.e., first wells have been drilled successfully for each). Each potential new oil field can be scored for its seismic upside potential, and reservoir simulation modeling can be used to evaluate its likely economic performance by scoring three functions within the suitability matrix:

- NPV of what is predicted from seismic monitoring and reservoir simulation
- Risk of further exploration success, in terms of the seismic data and the reservoir simulation of what was found
- Drilling success from the wildcat wells

The particular prospect shown in figure 7–1 has an NPV score of 5, an exploration success score of 5, and a drilling success score of 11. The weighted scores have been obtained by best-practices lessons learned from the scores of previous oil fields of this type that have been fully developed by the company. The scores from this field can now be quantitatively compared with other prospects from all over the world, to decide where next to invest new capital.

The *balanced scorecard* is another example of a way in which many companies improve corporate performance by using weights based on past experience, but these do not improve the linkages among humans, machines, and computer models in the unified, integrated manner of CALM. Nor does the balanced scorecard link the tactical and operational performance metrics to drive strategic resource allocation and direction down to the operational level capable of selecting among many disparate alternatives.

Business drivers / Technologies	NPV	Exploration success	Drilling success
Seismic monitoring	3	1	5
Reservoir simulation	2	4	6

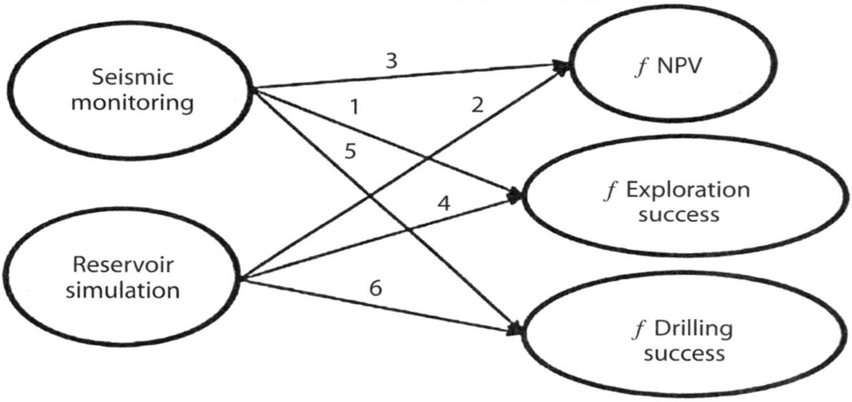

Fig. 7–1. Common exploration and production tasks, such as 4D seismic monitoring of real drainage patterns versus the reservoir simulations of that drainage can be scored using a suitability matrix. These tasks conform to CALM methodologies only after they are linked to NPV increases by way of modeled cause-and-effect scenarios that will lead to documented success. ***f*** represents functions of the inputs, such as summation. The larger the function, the better the score.

CALM implementations use chained matrices as a vehicle for mapping layered cause-and-effect relationships, like symptoms-to-problems and then problems-to-solutions among technologies, processes, and organizational boundaries that compose the many levels of every project. Symptoms can therefore be linked to problems, which can in turn be linked to solutions if the nature of the task is diagnostic. Then, metrics for how the solutions turned out can be gathered, so that cause and effect can be estimated. Finally, weighted scores of goodness of outcome can be summed across rows and columns to decide appropriate actions (fig. 7–2).

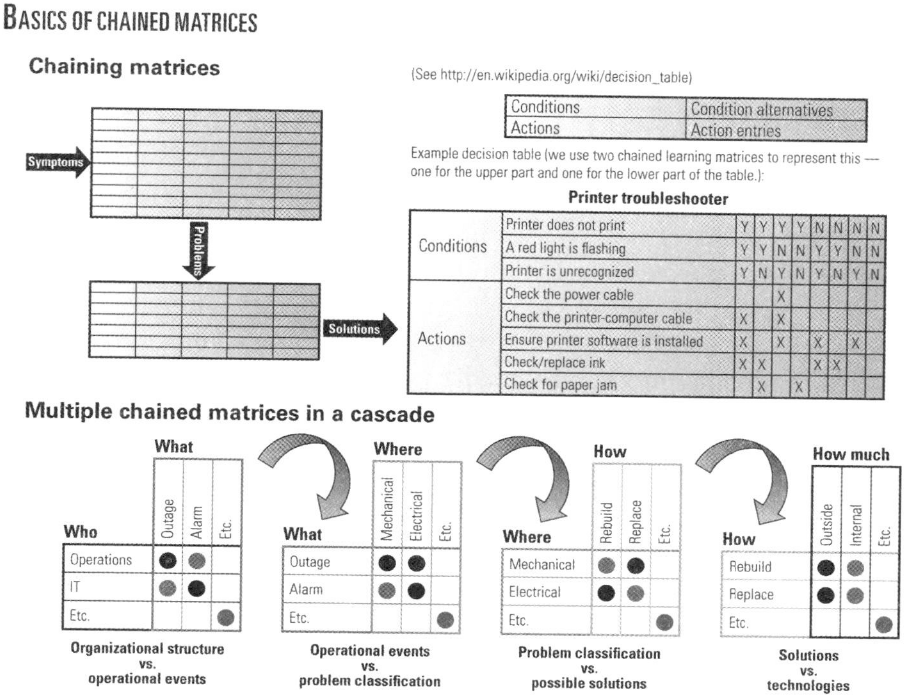

Basics of Chained Matrices

Chaining matrices

Symptoms
Problems
Solutions

(See http://en.wikipedia.org/wiki/decision_table)

Conditions	Condition alternatives
Actions	Action entries

Example decision table (we use two chained learning matrices to represent this — one for the upper part and one for the lower part of the table.):

Printer troubleshooter

Conditions	Printer does not print	Y	Y	Y	Y	N	N	N	N
	A red light is flashing	Y	Y	N	N	Y	Y	N	N
	Printer is unrecognized	Y	N	Y	N	Y	N	Y	N
Actions	Check the power cable			X					
	Check the printer-computer cable	X		X					
	Ensure printer software is installed	X		X		X		X	
	Check/replace ink	X	X			X	X		
	Check for paper jam		X		X				

Multiple chained matrices in a cascade

Who / What: Outage, Alarm, Etc. — Operations, IT, Etc.
Organizational structure vs. operational events

What / Where: Mechanical, Electrical, Etc. — Outage, Alarm, Etc.
Operational events vs. problem classification

Where / How: Rebuild, Replace, Etc. — Mechanical, Electrical, Etc.
Problem classification vs. possible solutions

How / How much: Outside, Internal, Etc. — Rebuild, Replace, Etc.
Solutions vs. technologies

Fig. 7–2. Chained matrices. The columns of the first matrix become the rows of the second, the columns of the second then become the rows of the third, and so on.

The methodology of chained matrices is to make the vertical columns of the first matrix become the horizontal rows of the second, and so on through the series. A user can enter the decision table by comparing who with what, then moving to the matrix that links what with where, and proceeding on to the matrix that links where with how, and, finally, asking how much. CALM adds to the use of chained matrices additional tools from statistics, stochastic control, real-option theory, and ML to build a decision support system that estimates the optimization of business and engineering objectives simultaneously, while also considering uncertainty.

TRANSPARENT PERFORMANCE METRICS

Performance metrics and their efficacy in tracking process improvement are critical to lean implementations. Normally, the many facets of a complex system will hinder the building of a single objective function to define optimization using performance metrics. If multi-objective tools are needed, then you can use Pareto efficient frontier curves for determining and visualizing the different trade-off options. A Pareto frontier is the surface of optimal trade-offs for the objective parameters that are most efficient. The Pareto efficient frontier in portfolio management is very effective at selecting which properties among many prospects to buy, sell, trade, or develop. Interacting with the trade-off curves determines the "geometry of your business" in terms of the balance between risk and reward. Visualizing and incorporating the Pareto efficient frontier into algorithms can help formulate robust strategies and solutions to operational decisions that have many conflicting considerations, such as component replacement strategies in preventive maintenance programs.

Consider the decisions that have to be made to select the right mix among many choices to replace or reinforce electrical feeder components in a grid or among many options in an old oil field in order to improve performance. In electrical component maintenance, the mix could include how many old cable types to take out, versus overloaded components to replace, versus strengthening with additional components. In oil and gas production, it could be the number of modern downhole pumps to add, versus surface pump jack conversions, versus new wells to drill.

Modeling (often generated by ML) allows the business to map out the parameter space to explore the trade-offs among cost, benefit, and the mix of what work to do. There are "sweet spots" in the Pareto efficient frontier that defines the business geometry (fig. 7–3). The topography of this optimal surface allows choices among several options along this surface, depending on your cost, benefit, and preferred asset-mix strategy. Your company may always want to operate at today's highest peak in potential benefit at reasonable cost (A in fig. 7–3). However, there may be other mixes of components on the surface that offer sufficient robustness to changes in exogenous variables, but at lower cost and acceptable benefit (B in fig. 7–3). Understanding the shape of these Pareto surfaces—again, this is the geometry of your business—can help decide actions to take to optimize operational performance.

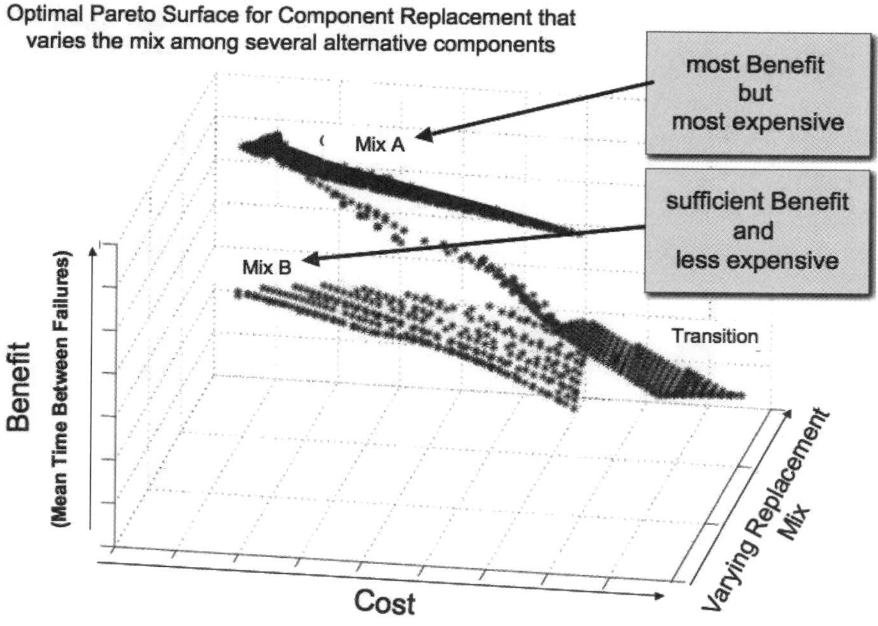

Fig. 7–3. Pareto optimal surface for deciding upon component replacement mix strategies. The 3D landscape allows decision-makers to see where the company is most comfortable in the trade-offs among benefit and cost for several different replacement mixes. (***Source:*** Adapted from C. A. Mattson and A. Messac, Pareto frontier based concept selection under uncertainty, with visualization, ***Optimization and Engineering,*** 6: 85–115, March 2005)

RL CONTROLLER

A key to lean implementations is to continually seek perfection in performance, improving through rigorous enforcement of feedback loops (termed *Kaizen* by Toyota). Within CALM, software automatically tracks operator actions, scores the outcomes of those actions, and then propagates corrections back to optimize performance through the feedback loops.

The adaptive learning algorithms of RL may one day close the loop to assemble a lean, gap-free integration of these core business processes. Once process improvement is decided upon, it can be measured. The RL controller also can find systemic synergies among project efforts by using simulations via the ISM. The challenge is that completing this feedback loop at the enterprise scale and carrying out the computations efficiently appears daunting today. It is feasible today with smaller control problems—such as power load transfer for a network, or optimizing the profitability of a pump jack or perhaps an oil field.

REAL-OPTIONS CAPABILITIES

Real options are used in industries, such as pharmaceuticals and aerospace, and on Wall Street as a guide to make quantitative evaluations of what is most likely to be useful to drive systems improvement throughout all life-cycle stages of a product. Decision and financial discipline must be developed and applied at all levels of the lean system in the energy industry as well. Parallel design evaluations then allow the real-option valuator to drive cross-system optimization. Processes can then be transferred *literally* to new development planning for the energy industry. Both the real-options framework and the adaptive RL controller use the same ADP algorithms as a means of evaluating future actions under uncertainty.

Scenarios also form a fundamental evaluation tool that complements real options. The volatilities of the stochastic inputs to the real-option evaluator can be set by the scenarios. The ISM provides the simulation environment to propose good and bad scenarios and test the predicted outcomes of each. These can be grouped as good or bad, expensive or cheap, and environmentally green or brown, for example.

PUTTING IT ALL TOGETHER

CALM has steadily evolved from process control to information management and then knowledge management. Automated, real-time optimization of the business is now becoming feasible, even though it wasn't practical just five years ago. Not only do lean systems tell the decision-maker what might happen next, but they also present contingency information about what might happen next, in a clear and concise way, at all times.

Operators need help particularly when multiple areas experience significant problems at the same time. Lean decision support provides not only what is likely to happen next but also what the risks and ramifications of different preventive remediation sequences that may be enacted are. Action tracking and key performance indicators record actions taken and then provide a basis for future decisions, by use of the RL controller, so that the system will get increasingly better at decision support.

As an important side benefit, especially with the average population of employees in both the oil and gas and electricity businesses aging fast, training becomes much more effective with these CALM tools. The RL controller can be used to advise or guide new operators on what actions to take in the same manner that car navigation systems guide drivers.

BOX 7–1. LEAN SUPPORT TOOLS

- Collaborative tools:
  - Value-added analysis (VAA)
  - Autonomation (the transfer of human expertise to machines, e.g., robots)
  - *Jidoka* (the famous shut-down option if anyone spots something out of specification)
  - *Kaizen* (continuous improvement)
  - *Kaikaku* (a Kaizen blitz that rapidly analyzes steps to take for continuous improvement)
  - Total production maintenance (TPM)
  - *Hoshin* (strategic planning)
- Just-in-time tools:
  - Just-in-time delivery (JIT)
  - *Kanban* (scheduling deliveries to have just the right amount of supplies on hand at any one time—no more, no less)
- Visibility tools:
  - Capital value process (CVP)
  - Assembly simulations (AS)
  - Value stream mapping (VSM)
  - Quality function deployment (QFD)

- Integration tools:
  - Supply-chain alignment
  - Workflow management systems (WMS)
  - Integrated master plan (IMP)
  - Integrated master schedule (IMS)
  - Integrated system model (ISM)
  - 3D solid models
  - Digital parts libraries
  - Program execution plans
  - Data warehousing and data cubes
  - Middleware like publish/subscribe and Web services
  - Wrappers for legacy code
  - Business process execution language (BPEL)
- Benchmarking tools:
  - Earned value management (EVM)
  - Economic value added (EVA)
  - Best practices
  - Real options (RO)
  - Sensitivity analysis
  - Flexible manufacturing
  - *Taguchi* (statistical quality measures)

- Control tools:
  - Electronic document management
  - Precedence diagrams
  - Product life-cycle process (PLCP)
  - Six Sigma
  - Root-cause analysis
  - Activity-based costing (ABC)
- Metrics tools:
  - Metrics thermostat (MT)
  - Dashboards and scorecards
  - Applied information economics (AIE)
  - Action trackers (ATs)
  - Quality, cost, delivery, safety, and morale (QCDSM)

8 OIL AND GAS OPERATIONS

The oil and gas business is the largest nonmilitary endeavor ever undertaken, and the profitable extraction of hydrocarbons in the future is critical to supply the world's ever-increasing energy demand, especially with China, India, Russia, and the rest of Asia developing at such a rapid rate. As wer go to press in 2008, the industry is estimated to be spending $1 trillion per year on oil and gas fields, pipelines, drilling rigs, and production platforms, $100 billion of which is just in the deepwater around the world, and those expenditures can only increase as the ultra-deepwater moves into full development. However, as we have seen so far, the very manufacturing process by which we produce oil and gas is in need of being made more efficient, to ensure economic success in not just these, the largest and most risky of all capital investment projects, but also in the other operations and construction projects that power this global business (fig. 8–1).

There are fundamental reasons why systemic improvement of the production process itself is a difficult mission for the oil and gas industry. Principal among them is the fact that the manufacturing process is about as difficult a "factory" environment as could be imagined. Other giant manufacturers around the globe also deal with multiple vendors, hundreds of thousands of parts, and millions of customers, but the location of manufacture of the energy product is *not* in a controlled factory environment. Instead, it is beneath deserts, glaciers, fields, cities, oceans, and lakes; wherever oil and gas happen to be discovered beneath the Earth's surface. Furthermore, the geological resources themselves are located up to five miles underground.

Design/Build of same size and magnitude as Space Station

AS IS:

- **Poor communication of data and design among subs**
- **Modifications and re-mods not controlled**
- **Sub-Assemblies not combined into Total structural solid**

TO BE:

- **Great communication of data and design to all**
- **Easy to understand the effect of modifications**
- **A true Digital Manufacturing and Production System**

Fig. 8–1. Structures built for the energy industry—every bit as large and complex as a space station. However, the as-is condition of that construction is not as good as that of the aerospace industry. (***Source:*** R. Anderson and W. Esser, Ultra-deep E&P: How to operate the advanced digital enterprise, ***Offshore,*** October 2000)

Since the production of oil and gas is a huge optimization problem, it provides an excellent example of the results that can be realized from the implementation of CALM in the field-oriented energy industry. Processes and economic models must be constructed that integrate the vast stages of development—from appraisal, through planning, construction, and operations, to abandonment—with all levels in the organization, from planning through scheduling, supervisory control, regulation, reservoir management, engineering decision-making, and environmental impact involved (fig. 8–2). In this chapter, we will give examples of oil and gas implementations that walk through the systems integration requirements of CALM.

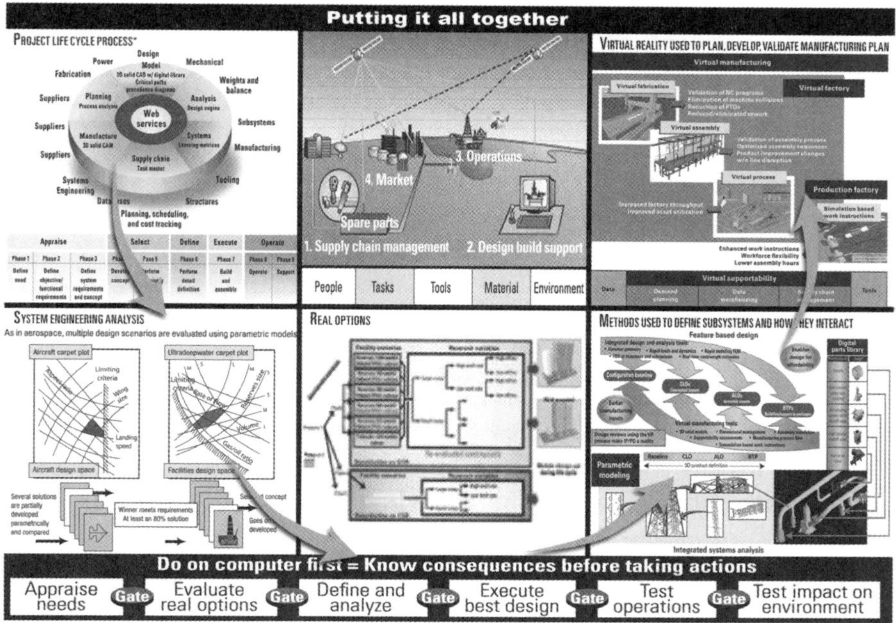

Fig. 8–2. The system engineering components of the CALM methodology—using PLCM (upper left) to sweep from systems engineering analysis (lower left) to real-options evaluation (lower center), subsystems definition (lower right), and validation by use of visualization (upper right) to integrate an entire project (center). Details of each panel are explained in chapter 4.

Industries such as automotive, aerospace, pharmaceuticals, and the military have cut capital expenses, operating expenses, and cycle times up to 25% yearly through lean practices. We can expect a similar paradigm shift in the oil and gas industry in that CALM systems should favorably affect cash flow and profitability of energy development projects to the same degree. For example, CALM requires significantly more integration of software tools than is currently typical in oil companies. That way, operational and communications gaps are filled, and connectivity is maintained among a large number of contracted partners. Such efficiency is necessary to maintain profitability in the petroleum industry in the face of the large up-front commitments of cash required before the first cash is seen from large oil and gas investments, whether they be oil and gas fields, liquefied natural gas (LNG) pipelines, or refineries.

The economic risk begins with the setting of investment and profitability goals by top management (fig. 8–3). These goals are calculated on the basis of the geological model of oil and gas production over time that is specific to each reservoir. Huge amounts of capital are sunk into getting each asset into production (phase 1), then new cash piles in as production ramps up to a maximum (phase 2), but every year after that peak at the field depletes and slowly the cash flow dries up (phase 3) (fig. 8–3). Profitability continues long after peak production is realized because production and reservoir engineers are given strong incentives to maximize short-term cash flow. Therefore, they engage a vast array of enhanced production tools and techniques to sustain production volumes. Then, the inevitable decline in production overwhelms even the best of oil field technologies, and profitability is lost even while costs are cut more and more each subsequent year until abandonment.

ECONOMIC PERFORMANCE OF EVERY FIELD HAS THREE PROFITABILITY STAGES

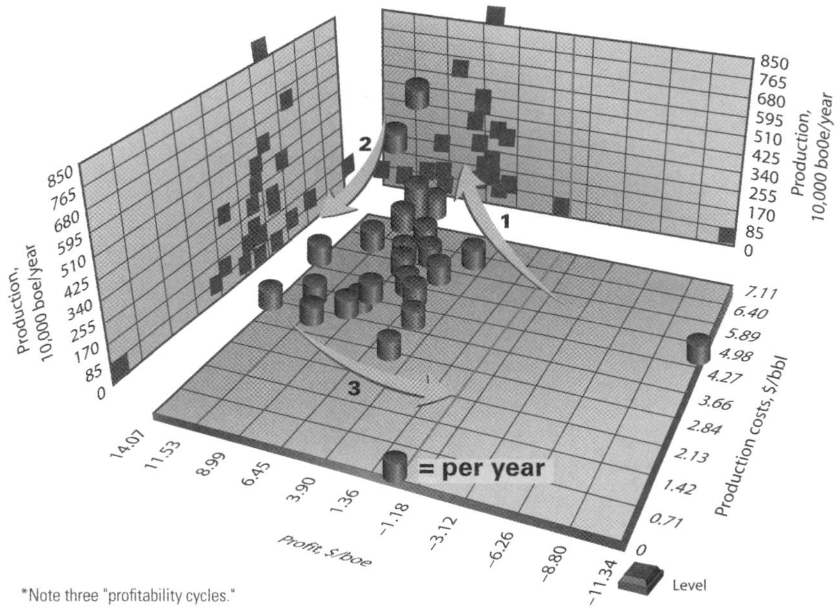

Fig. 8–3. The year-by-year profitability (cylinders) of a typical oil and gas field, which proceeds in three phases during the life cycle. Profit and costs vary in a systematic pattern as production increases from zero to peak (phase 1), then holds steady as enhanced production technologies hold production at a plateau (phase 2); inevitably, though, geological depletion produces declines that drop year in and year out until abandonment (phase 3). 3D locations within the cube are projected as 2D shadows on the backplanes. Data are a snapshot of oil production from 24 oil and gas fields in various stages of depletion from the South Eugene Island and Green Canyon areas of the offshore Gulf of Mexico in 1995.

In fact, it is the connectivity between the design, build, and operate silos that provides the greatest economic opportunity for enhanced benefits from CALM. This linkage drives a need for the upstream oil and gas industry, in particular, to adapt lean systems integration techniques from other industries—and, interestingly, from their own lean and efficient downstream—even though upstream margins are often more than 30% ROCE and downstream <5%.

In the downstream refining and marketing business, margins are so slim that process engineering has been carried to a high art. Systems are integrated, plants are managed with state-of-the-art process management software systems, and personnel are mostly cognizant of the profitability of their decisions. Total plant integration of operations is preached throughout the ranks of the company. This is a necessity when margins hover around 5% ROCE or less. In the following sections, we will present examples of the application of CALM techniques to increase profitability in petroleum industry environments.

EXPLORATION AND PRODUCTION

An oil company must not only deal with a variably priced commodity but also develop their "factories" in remote, nonoptimal locations, with each customized and configured as a one-off facility. Also, each asset manager must plan for operations that will be continuous and, hopefully, efficient for at least 30 years after first oil and gas is delivered to the surface. These field facilities must then be integrated into an interconnected chain with other factories, all tied by pipelines and seaborne train and truck traffic to trading floors, refineries, and finally customers that often live halfway around the Earth (fig. 8–4). Consequently, oil companies must continually reevaluate the assumptions behind all aspects of how they manufacture oil and gas—particularly in the hyper-expensive ultra-deepwater offshore technology industry. This includes reexamination of traditional approaches to manufacturing and operations, which, to date, have been dominated by high-redundancy, overdesigned and overbuilt facilities

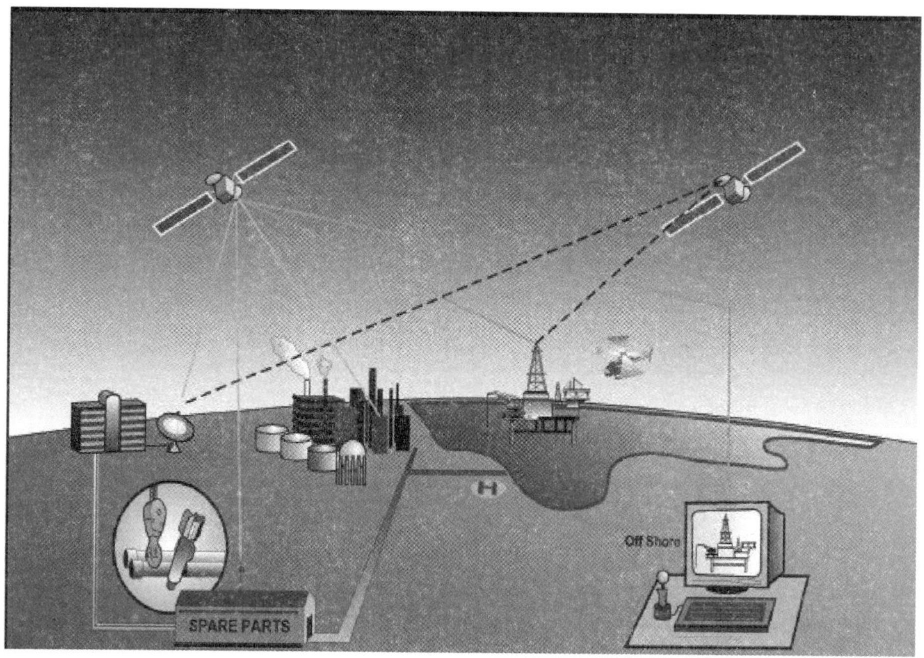

Fig. 8–4. Oil and gas production—a sprawling enterprise that requires integration into the larger design/build/support infrastructure of the overall industry manufacturing process

Increasing productivity

Because of the tremendous scope of the ultra-deepwater offshore industry, technological improvements that create even a small percentage increase in productivity have a significant impact in absolute dollar terms on the profitability of even the largest of the supermajor oil companies. Therefore, the successful implementation of more efficient lean manufacturing processes can result in significantly increased profit margins. The success or failure of some of the world's largest oil companies depends on the improvement of their ultra-deepwater enterprises.

The best-performing companies in the energy industry have been as well schooled in process improvement as companies in any manufacturing industry. They can properly frame their problems, gather the facts that impact their decisions, decide quantitatively and decisively, and monitor and react to the probabilities and contingencies of those decisions.

What even the best energy companies are trying to improve, however, is the integrated system itself: the enterprise-wide design/build/support system-of-systems necessary for game-changing improvement of the oil and gas manufacturing process. CALM asset tracking, logistics warehousing, computer-aided manufacturing, business simulation, and optimization loops, which are common in "easier" manufacturing industries (e.g., automotive, aerospace, and pharmaceuticals), have not yet been deployed widely, particularly in the offshore industry.

Individual oil and gas capital projects can range in cost from $10 million to $5 billion—and sometimes even more. In the future, they must take on the added burden of becoming environmentally green oil fields. Environmental stewardship and cleanup requirements create the opportunity to apply CALM principles to develop cleaner design/build/support structures for the life of the asset processes. Currently, these are focused on reducing the time and the cost of the design-to-produce cycle. Cutting these costs by half would quadruple a company's profits from oil and gas projects and would provide ample cash to pay for greenhouse gas and other environmental improvements. Dave Lawrence, then global exploration manager for Royal Dutch Shell, made the following comment during his keynote address to the World Petroleum Congress in 2000:

Shell approaches deepwater development on something of the same level that NASA approaches space projects. Deepwater projects are as remote as the moon, and the costs are about as much. Like outer space, exploration in deepwater poses huge risks, where even a minor glitch can be disastrous. It therefore requires detailed, advanced planning driven by the development of new technologies. Shell is committed to a reduction of deepwater development time and cost by 40% each, while at the same time increasing the environmental and safety programs to ensure fewer accidents.

Better supply-chain management

The as-is situation will be difficult to correct quickly, because most oil fields are one-off designs. There have never been true systems integrators to enforce commonality as there are with Boeing and Lockheed Martin in aerospace. The subsurface geology of oil and gas deposits is truly different from field to field, and it changes even from the beginning to the end of production in each oil field. Unquestionably, innovation in the industry is being driven by the developments in the "ultra-deepwater triangle"—the Gulf of Mexico to offshore Brazil and West Africa (from Nigeria south to Angola). Companies operating in these harsh conditions are required to produce oil and gas in water depths ranging from 6,000 to greater than 10,000 feet—and that just gets the drill bit to the seafloor. Companies operating in these harsh conditions are setting the new standards for the industry in the design of all aspects of the oil field of the future.

Concerning the to-be state of the future oil and gas production facility, the industry should be searching far and wide for more cost-efficient processes. From subsurface factories to seafloor completions, the integration of these modern systems into overall life-cycle operations is at the top of the development list, even at $130 oil and $11 gas. Implementation of CALM tools and processes, such as those used by Boeing, for the design and support of offshore facilities will lead to revolutionary cost and cycle-time savings; they certainly have in the aerospace business. Interestingly, expenditures comparable in size to capital design/build budgets in the energy business were required to develop these lean aerospace tools and processes into successful products and services.

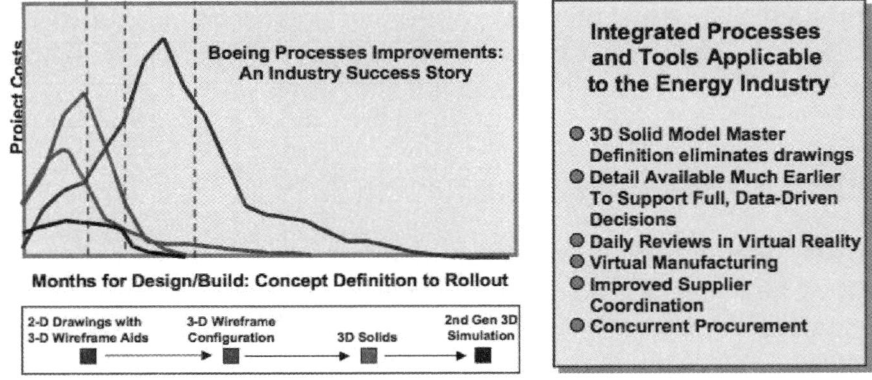

Fig. 8–5. Industry-to-industry best practices—creating game-changing opportunities. Dramatic cost reductions were experienced at Boeing when they converted to CALM methodologies. (***Source:*** Anderson and Esser, ***Offshore,*** October 2000)

The average development time from discovery to first oil production in giant oil fields of the world is currently five to seven years and is as much dependent on the availability of infrastructure and support services as on the development of new equipment and technology. As techniques for CALM operations are proved to be successful, this cycle time will likely be reduced to three to four years. For example, operators should be seeking to reduce the developmental time by standardizing the size and complexity of deepwater structures in order to minimize the fabrication time.

Boeing has accumulated economic metrics that define the value of its four generations of digital enterprise improvements since the 777 aircraft (fig. 8–5). These cost and cycle-time reductions represent significant value if they translate into similar savings in the production of multibillion-dollar, ultra-deepwater oil and gas facilities.

The as-is supply chain in the upstream oil industry is classic "bricks-and-mortar," having been built up over the past 100 years and composed mostly of company-specific inventory networks. That is, individual oil and service companies are basically on their own to provide spare parts and maintenance for their assets in oil fields around the world. There is minimal sharing of assets and resources, as well as very little sharing of production standards, let alone inventories.

Industry-wide supply-chain transactions involving original equipment manufacturer (OEM) spare parts and maintenance are estimated to be a hefty $60 billion per year. Of that, $25 billion is spent on expendables. There are more than 27,000 suppliers of OEM parts to the oil industry, but 90% are supplied by only 2,000 of these companies. Of oil companies' operating expenses, up to 10% is spent on expendable supplies used routinely during operations, and approximately 20% each is spent on rotating parts and equipment, consumable spare parts, and overhauls. Thus, the supply chain must be considered ripe for CALM savings in any to-be redesign of production methodologies.

Internet businesses are popping up all over the oil patch, as with many dealing with supply-chain management. Many of these portals have plans to expand into virtual inventory management. We have encountered no company in the current electronic marketplace, however, that is planning to link these to their own real warehousing and distribution networks. Linkage of the electronic world to real warehousing and delivery (through the last mile) has proved to be the critical game-changer that produces lean efficiencies in other industries.

So far in the oil industry, we find no inventory management service of the size and scope of Boeing's Global Airline Inventory Network (GAIN), for example. Boeing's metrics for GAIN include 5%–15% price reduction in spare parts and supplies, owing mainly to reduced maverick purchases, increased usage of preferred suppliers, and improved leverage in contract negotiations through an online e-commerce portal. This resulted in a 35%–50% reduction in costs to store and turn the inventory (the average turn rate for a CALM supplier increased from less than once per year to more than four times per year in aerospace). There was also a 70% reduction in administrative costs associated with the overall supply chain. Each 1% reduction in industry-wide, supply-chain costs would add more than a billion dollars to the bottom line annually to any oil company of any significant international breadth.

GAIN is a lean, electronic, supply-chain model that should be transferable to any heavy industry manufacturing process, such as that of oil and gas development. The basic concept is that GAIN allows the consolidation of physical inventories from many customers by making them available through a global inventory management system. By aggregating customers' inventories, Boeing can force the end-to-end supply chain to greater efficiencies than an individual customer could provide on its own. Consequently, better service levels can be provided with half the inventory (fig. 8–6).

CALM services like GAIN can produce optimization of the manufacturing process involved in the upstream oil and gas business as well, by driving standardization of equipment used in exploration and production activities through communal supply-chain management. The core of the GAIN paradigm for the oil and gas industry would be a worldwide inventory management system for the design, build, operation, and maintenance of production facilities (fig. 8–7).

One goal would be to improve the system of oil and gas production by importing new CALM technologies and techniques from such far-field industries that are much better at it than the oil industry is. Another goal would be to address system bottlenecks and redundancies, with a principal focus on advanced information and communications technologies worldwide. We are certain that the application of such enterprise-wide lean tools with enterprise-deep lean services will dramatically reduce the time and the costs associated with offshore oil and gas production infrastructure.

The lean combination of supply-chain management, data transparency, and efficient management processes produces concrete savings that come in two waves. The initial conversion to lean processes results in increased labor productivity, faster production throughput times, faster turn rates on inventory, fewer errors in final deployment, much less scrap material, and faster time to market (fig. 8–8). Continuous improvement produces another round of savings and increased profitability, because the delivery of each project is quicker than the previous one, just as in aerospace.

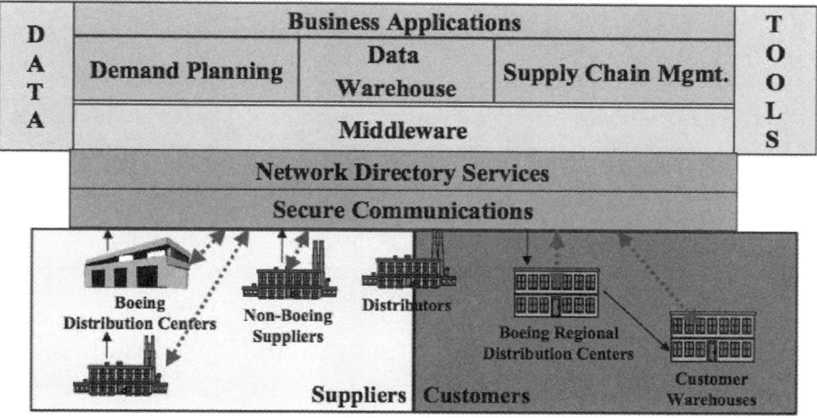

Fig. 8–6. Boeing's GAIN architecture, combining the information hierarchy (top) with the physical inventory consolidation and distribution system (bottom). Black arrows represent traffic in primary parts and supplies, and dotted gray arrows represent the information flow.

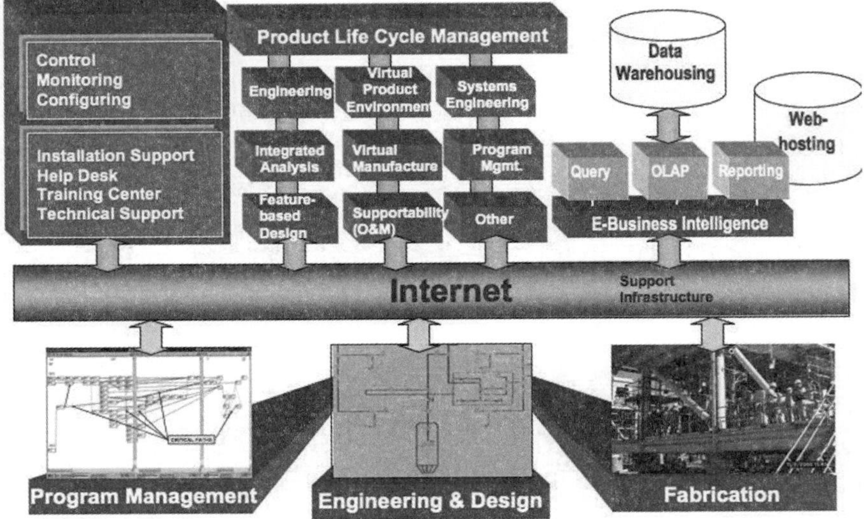

Fig. 8–7. GAIN's functional architecture—managing the supply chain for large capital construction projects, such as oil and gas production facilities. OLAP: online analytical processing.

In sum, Boeing has developed metrics and has documented the following lean improvements that should directly translate to the offshore technology industry using them:

- Cycle time was cut in half.
- Person-hours (including worker "touch" hours) were cut in half.
- Prototypes were made without one-off tooling.
- Physical mock-ups were eliminated.
- Engineering changes were cut in half.
- Paperwork really was almost entirely eliminated.
- Nonconformance-to-specification costs were reduced by 70%.

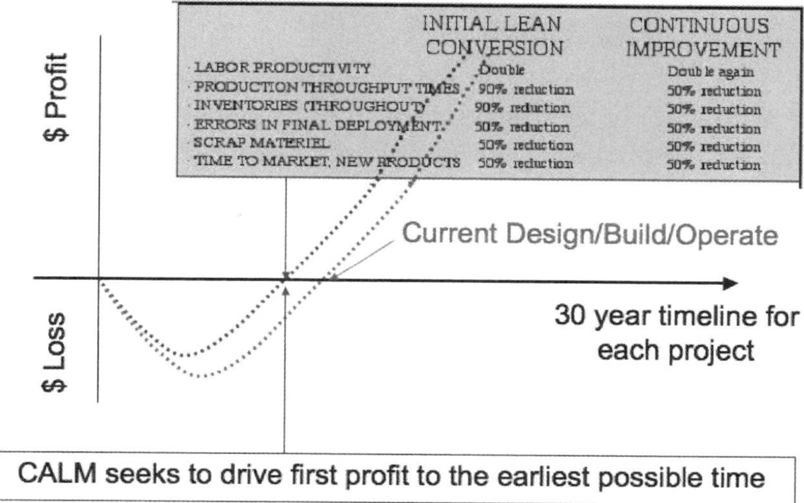

Fig. 8–8. Cost and efficiency savings, offsetting the risk of losses from the early commitments of cash that are required of all capital construction projects in the oil industry. Compare the shorter loss time from CALM projects (dark gray upper dots) versus that with current methodologies (light gray lower dots).

- Support labor was reduced by 80%.
- Inventory was reduced by 90%.
- Fabrication costs were cut in half, because lead time for component delivery throughout the supply chain was reduced by 70%.

PRODUCTION MONITORING

CALM attempts to establish a feedback loop between actions taken in the field and metrics returned that score the success or failure of those actions. This feedback is input into a model that continuously recomputes optimal solutions to keep the next actions always better than the last—thus, the conveyor belt of figure 8–9.

The tracking of oil and gas drainage in the subsurface over time (called *4D*) is a lean, time-lapse technique aimed at improving geological reservoir performance and production efficiency. 4D has introduced several powerful CALM tools to the development-engineering arsenal, such as 4D time-lapse seismic differencing, fiber-optic seismic monitoring arrays in wells, and downhole sensors of many other types such as pressure and flow rate. 4D holds great promise as the keystone of a new, integrated reservoir management strategy that is able to image changes not only within a reservoir but also within the stack of reservoirs that make up most of the oil and gas fields in the world today. 4D seismic monitoring of reservoir performance provides an excellent example of the power of such CALM systems—efficiently maximizing profitability even with extreme price fluctuations and geological uncertainties.

ISM

The oil and gas industry is only just now developing the infrastructure and controller logic for many components of 4D monitoring using an ISM. For example, 4D seismic monitoring is still centered on reacquisition using 3D methodologies that are hard to duplicate exactly. Consequently, field operators concentrate on seismic reprocessing and reinterpretation, instead of the differencing of time-lapse data that will show where oil and gas have been left behind.

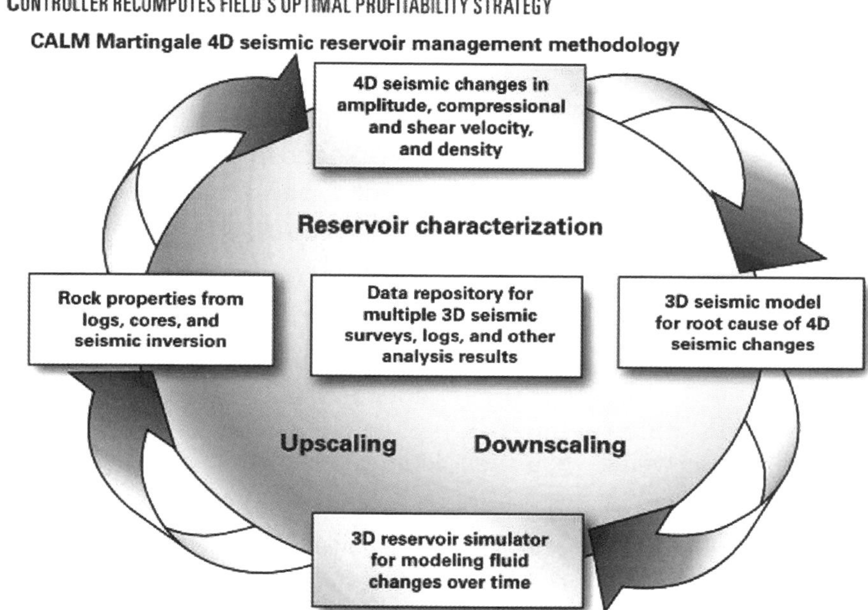

Fig. 8–9. CALM's never-ending feedback loop, providing a continuous conveyor belt made up of technologies that are constantly applied and reapplied to maximize profitability for an oil and gas production facility

A CALM architecture for 4D seismic reservoir management, in contrast, must be focused on the need to maximize profitability at all times and under all circumstances. The computer architecture that is required will integrate the observed 4D seismic differences with a continuously running reservoir simulation PM, to illuminate the production pathways of fluid withdrawal in each field. What is missing from most smart electronic (e-field) architectures is the computational operating framework (OF) that allows for seamless, rapid feedback between and among the many and varied software applications and data streams that are required for modern oil and gas reservoir management. This seismic/reservoir modeling is integrated into the CALM software suite as what we call a *4D Seismic Reservoir* (4D SeisRes) architecture.

One problem that produces errors in interpretation is that seismic modeling is 1D and 2D, rather than 3D (like Earth). To compound the problem, current analyses are built around one reservoir at a time, instead of the full Earth volume involved, in which stacked reservoirs perform as an integrated whole underground.

4D SeisRes requires the following workflow combinations:

- 4D analysis of time-lapse seismic changes for at least two (but up to as many as are practical) 3D seismic volumes acquired at different times during the production history of a field, as well as their time-depth conversion, normalization, and differencing.
- Well log analysis, looking for time-lapse changes in the same type of logs run over several different time intervals in wells and their depth-time conversion.
- Reservoir characterization of stacked reservoirs, using geo-statistical co-kriging (a technique to interpolate a weighted average of neighboring samples to estimate unknown values).
- Exporting of all data, with time stamps, into the same Earth volume model.
- 3D fluid-flow simulations, preferably using finite element modeling.
- 3D seismic modeling, preferably elastic, rather than acoustic, to generate synthetic seismic cubes matching the 4D seismic observation time steps.
- Differencing of 4D model versus 4D observed seismic data.
- Analysis of the difference-of-the-differences between the model and observed results.

- Optimization that identifies the necessary changes in physical properties of the reservoirs to match fluid withdrawal, pressure changes, and seismic differences as closely as possible.
- A continuous feedback loop to the beginning, so that the computation is dynamic (24/7) and is never completed until the field is abandoned.

The architecture and the design of 3D seismic modeling—to interact with statistical reservoir characterization, the 4D observed seismic differences, a finite element reservoir simulator, and seismic inversion and migration codes—constitute a sophisticated computational task that is not currently commonplace. This task requires an extensible OF that enables the interpretation workflow to move easily among various vendor applications needed to complete the feedback loop described earlier, from geological to geophysical interpretations and on to the engineering implementations required for optimal performance by the modern e-field IPT.

SEVERAL VENDOR APPLICATIONS ARE MANAGED SIMULTANEOUSLY

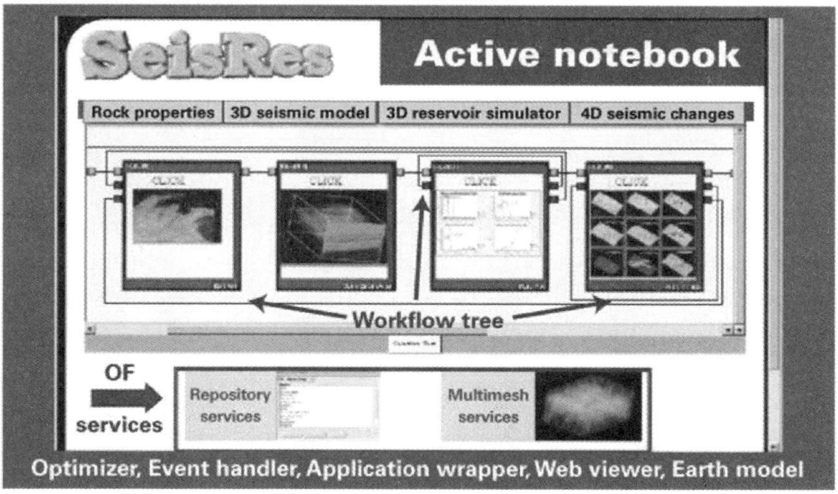

Fig. 8–10. The Web interface of the 4D SeisRes application. The feedback loop from fig. 8–9 appears in the connections between application wrappers in the workflow tree. Services are tracked at the bottom of the Web page.

PM

4D SeisRes design and implementation efforts are sizable and require the creation of IPTs. The architecture of the OF for the system that we implemented for an industry-sponsored consortium of oil companies contains the following major geological, geophysical, and engineering components (fig. 8–10).

- All actions within any of the many applications of the subsystems are recorded within the OF by use of a Web-based action tracker, which we call the *active notebook.* Like a research scientist's laboratory notebook, it allows any previous experiment to be redone.
- A vendor-neutral data model with a persistent input/output (PIO) data repository is required because the data sources for the required applications are not likely to be kept in the same data management systems. Present and previous analyses can thus be quickly and easily reviewed using the versions stored in the PIO data repository.
- Access to many vendors' applications is provided using automated *wrappers* (fig. 8–10). For example, the user can run popular reservoir simulators from Schlumberger (Eclipse), Halliburton (VIP), and other commercial models within the OF. The wrappers enclose the vendor applications and automatically manage the connectivity, data trafficking, and versioning of inputs and outputs.
- We add an event-handling mechanism to make applications run asynchronously (i.e., in parallel, but not necessarily at the same time). We parse the workflow among several applications simultaneously and distribute it to the client/server network, then reassemble it as completed. That way, we do not have to wait for one program to finish before beginning another.

- There must also be a rich set of reusable, extensible containers to hold engineering, geological, and geophysical data, so that new applications and data types can be added to the OF management system easily and quickly. Containers act as receptacles for objects with their associated data. Objects then can be swapped in and out of containers with ease.
- A 3D Earth model is next created (fig. 8–11).

HORIZON TOPS/BOTTOM MAPPED (LEFT) AND RENDERED IN EARTH MODEL (RIGHT)

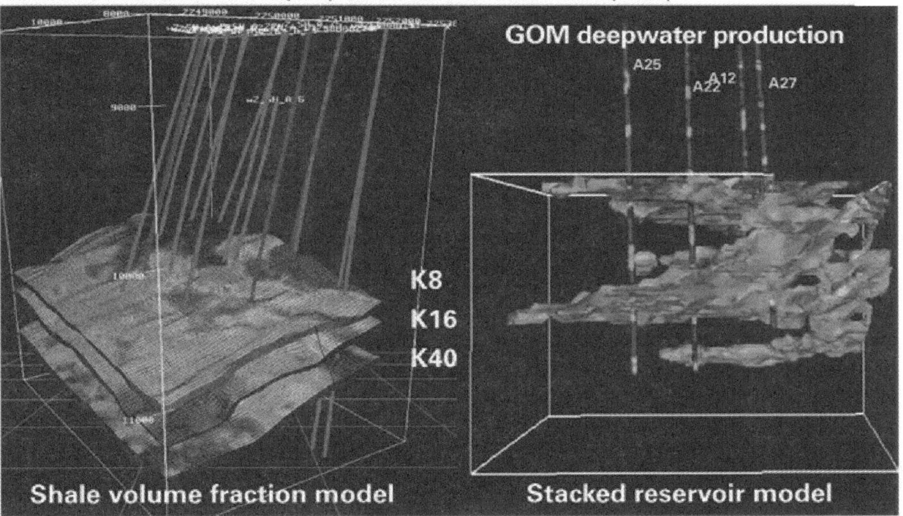

Fig. 8–11. Computer visualization of three oil reservoir tops and bottoms, the K8, K16, and K40 oil reservoirs (left), input into and rendered in 3D in the Earth model (right). Production wells (A12, A22, A25, and A27 wells) travel from the surface of the Gulf of Mexico (GOM) to tap the underground reservoirs.

- An automated meshing system is provided to create the framework needed by the reservoir simulator to build its model from any set of stacked horizons or other geological interpretations, such as scanned maps and charts (fig. 8–12). We use MultiMesh, an automatic meshing system from IBM Research that is topological, so that whatever the grid requirements of an application are, its 3D connectivity can be quickly computed in finite-difference, finite-element, upscaled, or downscaled versions as needed.

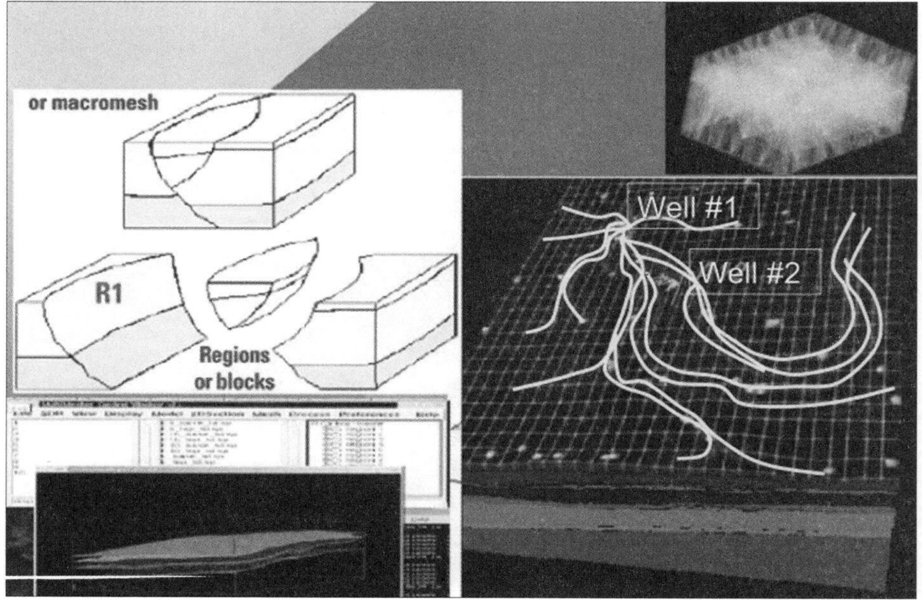

Fig. 8–12. The IBM MultiMesh system—creating a 3D mesh (top, right) from 2D horizons and tops (bottom, left). Like Legos, topological pieces of the complex oil reservoir can be taken apart and modeled separately (left) to simulate oil flow (streamlines) to well A12 (lower right)(see fig. 8–14).

- There is a common data viewer, a Web-based system that visualizes all application outputs. Users can then communicate the visualization of progress of their computer simulations over the Web to colleagues around the world, who can manipulate the images in real time as they are recomputing.

The key to 4D SeisRes architecture is the optimization laboratory that contains a set of tools that can be deployed at any time within the OF to provide parametric optimization services. The optimizer is implemented as a loosely coupled, component-based system, because the need for parameter estimation varies from application to application and from vendor to vendor (fig. 8–13).

The optimization laboratory allows a selection of reservoir simulation options, including hybrids that combine algorithms of different types, to produce the most appropriate flow-rate solutions. The technical goal is to facilitate the optimization process for the 4D SeisRes simulation.

The optimizer consists of three components—optimization solvers, forward simulation wrappers, and simulation data converters—each of which is developed separately for reservoir property characterization, reservoir simulation, petrophysical property characterization, and 3D seismic simulation. The optimizer converges on a single best-approximation result that simultaneously solves for permeability variations and flow-rate changes, with error estimators for each (fig. 8–13).

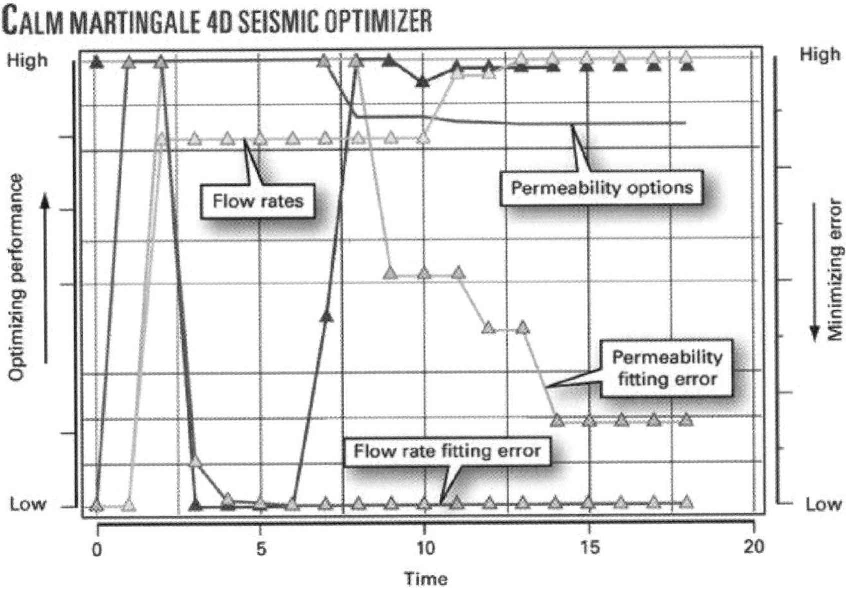

Fig. 8–13. 4D SeisRes optimization laboratory optimizes oil production performance by minimizing errors in reservoir rock permeability in reservoir simulation scenarios in order to maximize and sustain oil flow rates. Time scale is in months. Permeability is varied from node to node of the reservoir multimesh to match flow rates of oil and gas by minimizing errors in predicted vs. observed production histories of the wells and changes in 4D seismic amplitude over the 18-month period.

Consider an example that tracks the drainage of oil by two wells over a 15-month period using 4D seismic imaging (fig. 8–14). These four snapshots of drainage patterns were made by the 4D SeisRes system from interpolations using the reservoir simulation model, calibrated to acoustic changes caused by the depletion of the oil by wells A12 and A22, as imaged by multiple 4D seismic surveys made over an 18-month period.

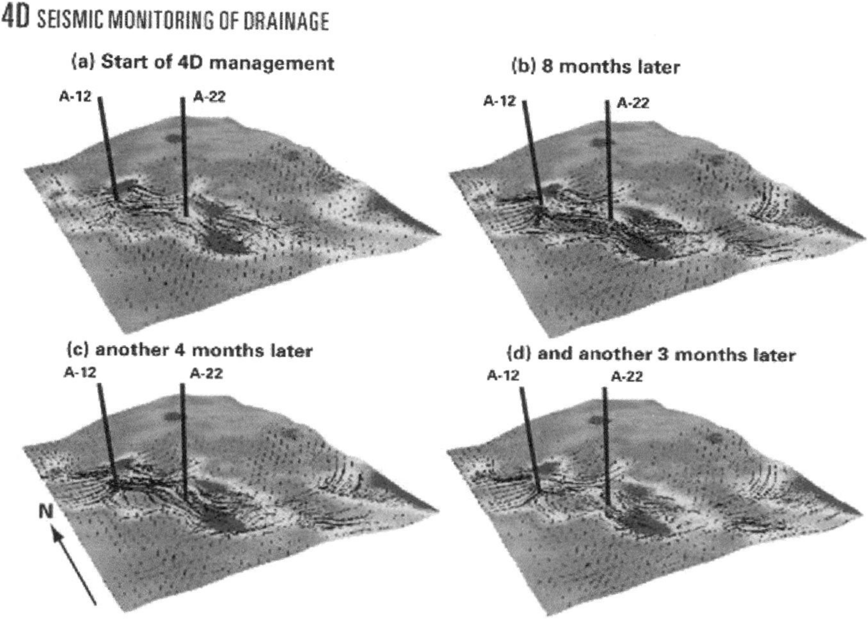

Fig. 8–14. Movement over time of oil (black streamlines) toward two wells (A12 and A22) plotted on a reservoir horizon. These four snapshots of drainage patterns were made by the 4D SeisRes system from interpolations using the reservoir simulator calibrated to 3D seismic surveys made at (a) and (d) 15 months apart.

A modern, real-time reservoir management system as a PM must be able to characterize multiple, sequential 4D seismic surveys; seismic attribute volumes that vary with offset; many repeated well logs of different types and vintages; statistically derived data volumes in spatial grids; 3D fluid-saturation volumes and fluid-flow maps from reservoir simulation; fluid-interface monitors; multiple horizons and fault surfaces; and so on. All this advanced geophysical technology is focused on visualizing remaining oil in reservoirs as it is swept toward the wells by pressure differentials set up by water sweeping the oil toward the surface (fig. 8–15). New wells were then drilled (white in fig. 8–15) where oil pay remains to be drained to the surface.

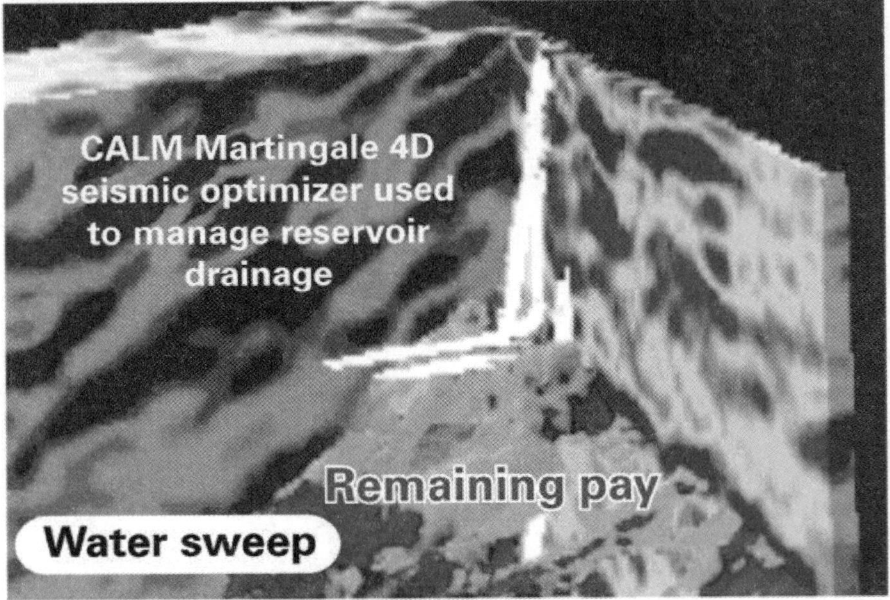

Fig. 8–15. Visualization of the 4D seismic image of a producing oil reservoir.

If the same CALM methodologies used in the design and build phases were employed in the operational phases of all fields, then this profit-to-loss cycle could be broken. In fact, lessons that petroleum companies have learned in the downstream systems point to successful ways out of this vicious cycle. PM architecture like 4D SeisRes can yield a global view of the complex subsurface reservoir to be achieved over time as oil is drained to the surface.

REFINERY IMPLEMENTATION

Batch runs of specific hydrocarbon distillation processes, such as making gasoline, jet fuel, and diesel from crude oil in a refinery, often deviate from planned specifications and schedules for many reasons including changes in market prices, consumer patterns, or late arrival of crude oil supplies. These changes from the scheduled work of the refinery have an impact on many areas outside the refinery, such as the spot prices and local shortages of gasoline, jet fuel, and heating oil.

Better scheduling of refining runs

In order to cut down on these disruptions in planned work, CALM techniques were introduced into operations of a large U.S. refinery. Issues related to fixing the timing and composition of mismatches of crude oil compositions relative to the demand for specific refinery batches were assessed by representatives of all the stakeholders involved, and a change-management plan was developed, communicated, and integrated into the production flow.

CALM methodology was used to identify the most common yet serious deviations from plan, by using a suitability matrix. Then the as-is stakeholder and scheduling relationships were mapped by BPM. Key team members were identified, and IPTs were created. Decision-making guidelines and performance metrics were established, critical timelines were assessed, and communications channels were formed. Policies were established for responses to deviations in the plan, and the to-be processes were planned, mapped, and changed.

The following CALM steps were executed during the implementation:

- Identify key stakeholders needs:
  - Any organization's success is a function of simultaneously satisfying needs of its stakeholders, employees, and customers.

- Establish stakeholder goals:
  - Employees: develop teams, particularly at the interface with management; provide business education and include everyone in face-to-face solution dialogs; create alignment/linkage to the business scorecard for everyone; have representation of all in the change project; align union and management vision within the personal career development process.
  - Owners: educate employees to become stakeholders with an owner mentality; align individual work to the top 25% of performance; set metrics; perform gap analysis; select big opportunities; reengineer key processes.
  - Customers: plan with them; focus work processes on their wants/needs; execute with operational excellence.
  - Vendors/suppliers: identify true interdependencies and transactional relationships; form partnering relationships for true interdependencies.

Combined change strategies were put in place at the refinery:

- *Executed an empowerment strategy:* provide face-to-face education and solution dialogs; understand the business side; establish lean linkage to business scorecard and first-quartile performance; provide business education; enable participant buy-in to change project; introduce culture of winning together; foster career development to increase individual contributions.
- *Followed a process improvement strategy:* identify and reengineer key processes; set up support systems with cross-functional teams; identify and implement technology enhancements.

- *Implemented the change plan:* keep units up if safe; design in reliability turnarounds; preventive maintenance; provide technical training; ensure adherence to operating guidelines; set operator rounds/inspections; protect quality; facilitate product specification management; implement laboratory monitoring; focus on the customer; build schedule and business plan with the customer/traders and suppliers; focus and align the core business; build integrated 60- and 15-day rolling schedules; communicate and align around enhanced 15-day schedule; execute for throughput efficiency; develop coordinated recovery plan for unplanned changes in schedule and/or specifications; manage product mix; develop economic models; identify cost of change from the basic crude oil mix.
- *Tracked the results.* The CALM implementation yielded the following successes:
  - Management reengineered to CALM processes: preventive maintenance; production structure; shift-to-shift communications; adherence to health, safety, and environmental guidelines all improved.
  - Successful cultural and leadership transformation: scheduling methodology for customer and suppliers; interdependent supplier relationships; partnering with subcontractors; joint recovery plans for future unplanned changes all established.
  - Reduced cost structure by an immediate 15%, with considerable upside potential for further improvement.

A key ingredient to the CALM success in this refinery example was the implementation of a total plant information system. During that implementation, the IT IPT accomplished the following on cost and on schedule, proving to skeptical management that they could listen to the customer and break down traditional IT silos (fig. 8–16).

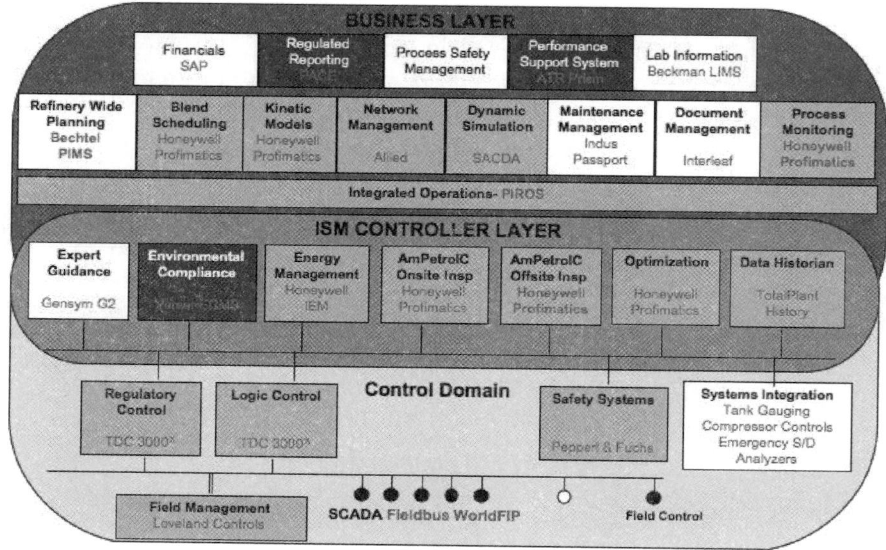

Fig. 8–16. Three CALM layers in the enterprise-wide and enterprise-deep information management system for the refinery "plant": information exchange domain (ISM Controller Layer); execution layer, where real physical and chemical processes were being controlled (Control Domain); and a newly integrated decision-maker business control domain (Business Layer)

IT improvements

The following steps were executed by the IT IPT:

1. Identification of all software systems controlling the various subsystems of the plant. A client/server networking datamart environment was designed to support integration into a central control facility, and a heads-up display system was built to show executives what is going on at all times in the plant.
2. Integration of all critical vender control software: Information management systems were designed and deployed for traffic (TIMS), batch runs (BIMS), logistics (LIMS), and real-time control (RTIMS). Communications middleware was developed

so that the 19 suppliers of the various software systems could transfer data, inputs, and outputs efficiently, and relevant statistics were integrated into an executive dashboard so that the decision-makers could monitor all critical software and information messaging traffic, to enable seeing the same information at the same time and place.

3. Installation of a total plant ISM-like system that included the following: logic for routing of blend schedules; systems for real-time monitoring of safety; subsystems integration of all physical plant monitoring; understandable human/machine interfaces; interdependent software supplier relationships and new partnering.

4. Results were tracked. The CALM implementation gave the refinery its first-ever, fully integrated, central control domain that linked all software and venders active in the plant. Results were scored and batch turnaround times were sped up by 15%–20%.

More than $200 million in additional profits were added to the bottom-line performance metrics of the plant over the next three years, and the change management plan became the benchmark for conversion of other refineries of the company to lean management worldwide. While this CALM success was being accomplished by one oil company, many others were cutting costs and eliminating maintenance and safety nets for their refineries, some with explosive consequences.

9 ELECTRIC OPERATIONS

Energy suppliers don't have the room to greatly increase the density of the electric grid or natural gas mains to accommodate continuingly increasing demand in major urban centers. At the same time, increased need for water, sewage, telecommunications, rail, and automobiles are pushing other infrastructure limits of cities worldwide. In particular, the giant economies of China, India, Japan, Korea, and the rest of Asia are at the forefront of this problem. European, North American, and South American cities are also challenged by enormous growth needs. Urban centers are all challenged to reliably supply the electricity and gas demand of today and accommodate the growth of tomorrow.

For example, the electrical infrastructure of New York City (NYC) currently has an operating margin of about 20% above peak consumption as of 2008, in order to provide high levels of service reliability. NYC's is the most reliable urban grid in the world. However, the pipes and wires of NYC are built to design standards that may not be sustainable in the future. Space is literally running out to install new cable ducts Maintaining the mostly underground electrical networks that make up the distribution grid is a further issue. Yet, utilities need to provide for growing populations and sustain or even increase current levels of reliability and customer service in the coming electric economy of the future.

Much of the physical expansion could be done in other ways. This transformation should be driven by CALM efficiencies, not by the sole addition of more copper, pipes, and other hardware. In fact, with strategic complications of new hardware and CALM, there will be cases where physical infrastructure could decrease.

In order to develop and deploy such intelligent efficiencies, energy infrastructure industries need to gain the business and operational capabilities to apply new lean technologies. In this chapter, we present examples of ongoing urban energy projects that illustrate the capabilities of CALM for increasing the intelligence of the utility business and the electricity grid. We will discuss the more general transformation to the intelligent grid that we predict will run the electric economy of the future in the final chapter.

SUSCEPTIBILITY TO FAILURE

CALM software tools, driven by novel ML algorithms, are presently providing real-time susceptibility rankings for primary distribution feeders in NYC that use SCADA data feeds and asset information such as installation dates, composition, and manufacturer to predict impending failures before they happen.

The mission of any control-center decision support system is to provide tools that will analyze risk for operators and suggest preventive maintenance, field actions, and policies to lower the occurrences of emergency incidents. Electric distribution failures put networks and field crews under considerable stress, especially during the high electrical loads of summer heat waves. Emergency failures cost hundreds of millions of dollars in operations and maintenance (O&M) expenses, not to mention increasing the risk of neighborhood electricity blackouts. Predicting impending failures so that preventive action can be taken before emergencies develop is a key to the evolution from reactive to condition-based maintenance of CALM.

IPTs of computer scientists and engineers from Columbia University and Con Edison have built a real-time, ML, contingency analysis system for control-centers that collects and analyzes real-time data thought to be relevant, along with component failure statistics like manufacturer, age, and past outage history, to predict future electric feeder failures in NYC. Included are shifted load support and feeder load ratings, as well as dynamic attributes like electric load and load pocket weights (a real-time measure of transformer banks that are off-line, open low-voltage electric mains and open distribution transformer switches). Power quality events of transients, over- and undervoltage swells and sags are also fed into the ML system to detect failure patterns and predict future susceptibility. Other CALM tools are currently providing new information about unexpected variances in the state of the cabe sections and joints, transformer loads, and other deviations from the normal behavior of the grid. The ML system also predicts rankings of the long-term survivability of the most-at-risk feeders for a capital asset prioritization tool (CAPT) to guide intelligent replacement and reinforcement decisions.

The data from static attributes that measure stress, wear and tear, and loading on the underground distribution feeders are combined with dynamic attributes from the continuous monitoring of load flow in all NYC networks to form the inputs into the ML system (fig. 9–1). A real-time ranking of the feeders is displayed that determines the most susceptible to impending failure. New rankings are computed every four hours and are fed to the contingency analysis tool (CAP), to estimate the risk of next-worst events that might happen. A priority list of the contingencies most likely to occur is pushed to control-center operators so that they can take mitigating actions.

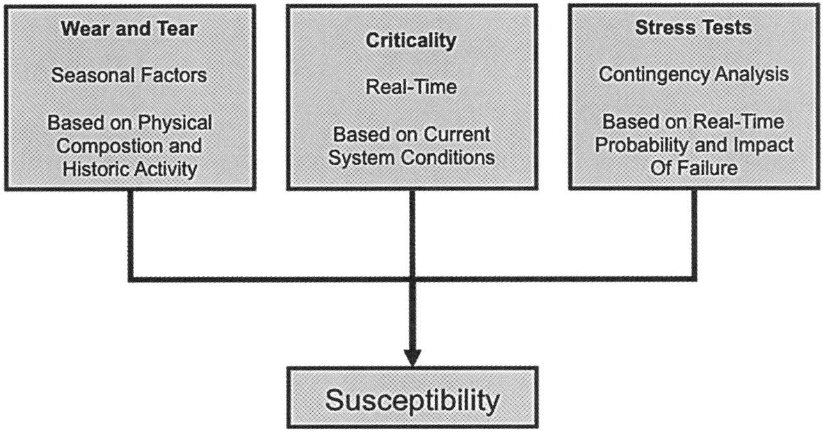

Fig. 9–1. Attributes that describe wear and tear (left), electrical criticality (center), and present state of stress of each feeder (right). These are fed into the ML ranking system to predict the feeders most susceptible to impending failure in a real-time contigency analysis tool called CAP.

There are now more than 400 dynamic and static attributes being fed 24/7 into the feeder susceptibility ranking system. The ML engine utilizes many different SVM and Boosting models to determine the most effective combination of attributes that explain past failures in order to predict the feeders most susceptible to future failure. These models are retrained every week. Then, a voting system scores the success of the prediction capabilities of each algorithm as real failures occur on the electric grid. The system reevaluates the state of the grid every four hours and re-sorts the electric feeders into a new ranking from most to least at risk. The CAP tool then combines this risk with the severity of the threats to advise operators of actions that can be taken to mitigate the risks.

We found that best performance was from an ML combination that includes SVM and a new Boosting algorithm for which we had developed the theoretical framework at Columbia but which we had not previously tested with real data (fig. 9–2). First, the feeders are sorted from those most to least at risk based on the single attribute that the ML system correlates with past failures, the history of past failures on each feeder. Then, a second round of sorting occurs, one with the most-at-risk half of the feeders and the other with the least-at-risk (best) half. The most-at-risk half are then sorted by the next-worst attribute of past failures. The best feeders, however, are sorted by the attributes that correlate with best past performance. Each succeeding round sorts a subset of the previous ranking until a stable success rate is determined by testing the ranking against the one-third of the data that were held out for verification. The overall ranking from the sort then predicts the success rate of future predictions (see http://www.phillong.info/publications/LS05_martiboost.pdf).

In the example in figure 9–2, a database from 2004 was used to train the algorithm for 2005 predictions. Incoming attributes were fed into two models: in the first, two-thirds of the feeder failures that occurred during the summer of 2004 were placed into a training data set; in the second, one-third were held out so that they could provide a blind testing data set. The testing data set was used to measure the predictive capacity of the training data set, and then the combined data were run through this model to predict feeder failures that had not yet occurred in 2005, a blind test.

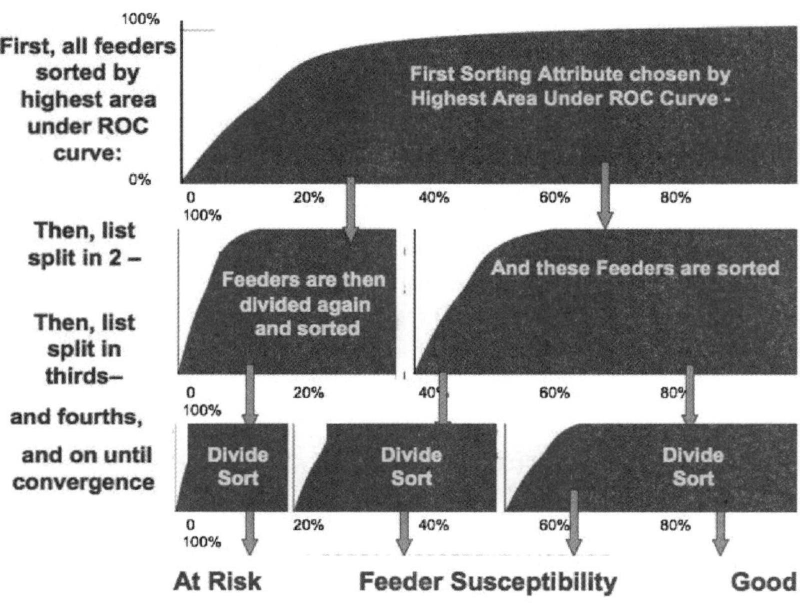

Fig. 9–2. MartiRank Boosting algorithm used to compute a ranking for each of the 400 attributes on the basis of that attribute's rate of success in predicting past feeder outages. Three rounds are illustrated to then sort them into most at risk to best rankings.

Importantly, the MartiRank Boosting algorithm does not win the voting from all training tests. The combination from the best-scoring algorithms is then sent as a susceptibility-to-impending-failure ranking to the real-time CAP decision support Web site for use by operations (fig. 9–3).

The ML system was developed to create real-time susceptibility feeder rankings in all boroughs of NYC. As with all true CALM systems, it tracks and scores its performance in real time so that the operator can tell when it is working well and when it is not. In this way, we have been able to detect seasonal changes that occur in the NYC electric grid. Summer predictions use a different mix of important attributes versus those in fall, winter, or spring. We discovered that this transition in system-states varies as the seasons wear on, hot to cold and cold to hot, and as the years vary from dry to wet, rainy to snowy, and so on, as the weather makes its indomitable effects felt across the city.

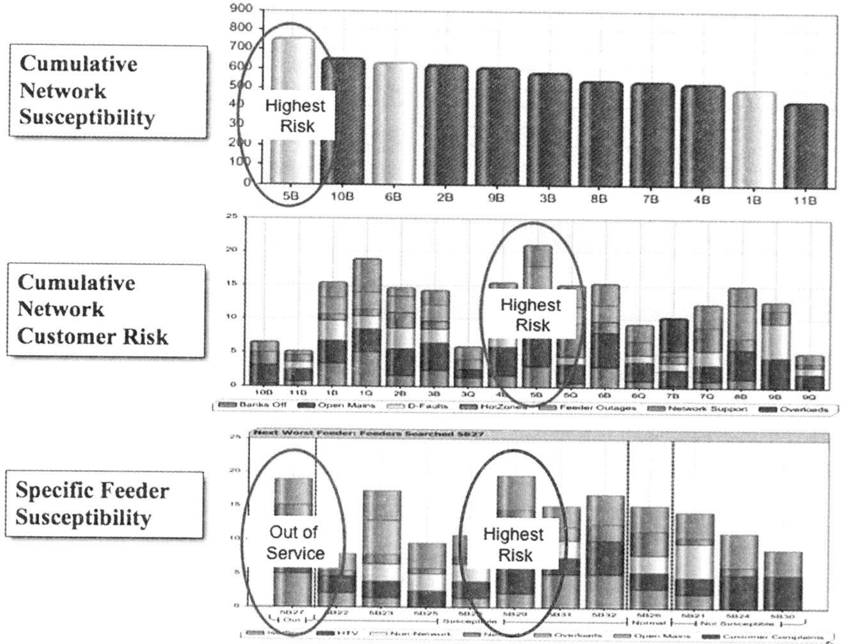

Fig. 9–3. Susceptibility-to-impending-failure ranking. The susceptibility ranking is pushed to the control-center operators as a contingency analysis of most susceptible feeders by network (top). Then, subdivisions of those most at risk for network failures are identified by heir risk to customer outages (middle). A feeder may be critical to the electrical continuity of a network but have little impact on customer outages. So those most susceptible for loss of customer electricity are also identified feeder-by-feeder within each network (bottom).

We have also incorporated the statistical concept of *control groups,* more familiar to the pharmaceutical industry. Feeders with similar key properties are selected as controls and contrasted with feeders that failed during the ML training period. Control groups are a powerful tool for determining the statistically likelihood of failure analyses of any kind, whether developing a new jet engine, curing a patient with cancer by using new therapies, testing a new AIDS drug, or as in our case, discovering how to prevent feeder failures in the electric distribution grid.

Specifically, the ML algorithms train themselves by comparing and contrasting attributes for each feeder that failed with attribute patterns from the control groups, selected first from within the same network and then within other networks from the same borough of NYC. Eventually, the ML system will also compare the attributes of the feeder at the time of failure with performance of the same feeder before and after it was repaired, looking for variances related to dynamically changing root causes.

We experimented with how often to retrain, and daily training gave the best performance over the summer of 2007. The susceptibility ranking was then updated every four hours by recomputing the attribute scores as load, load pockets, and power quality changed over the dynamic grid. Performance metrics that compare the blind test results against random guessing at the time of each outage are continuously displayed online.

The predicted importance of individual attributes to feeder failure is also recorded, so that when a failure occurs on one of our susceptible feeders, the five attributes with highest performance scores can be examined to better understand threats from daily variations in component configurations to the overall system behavior (fig. 9–4).

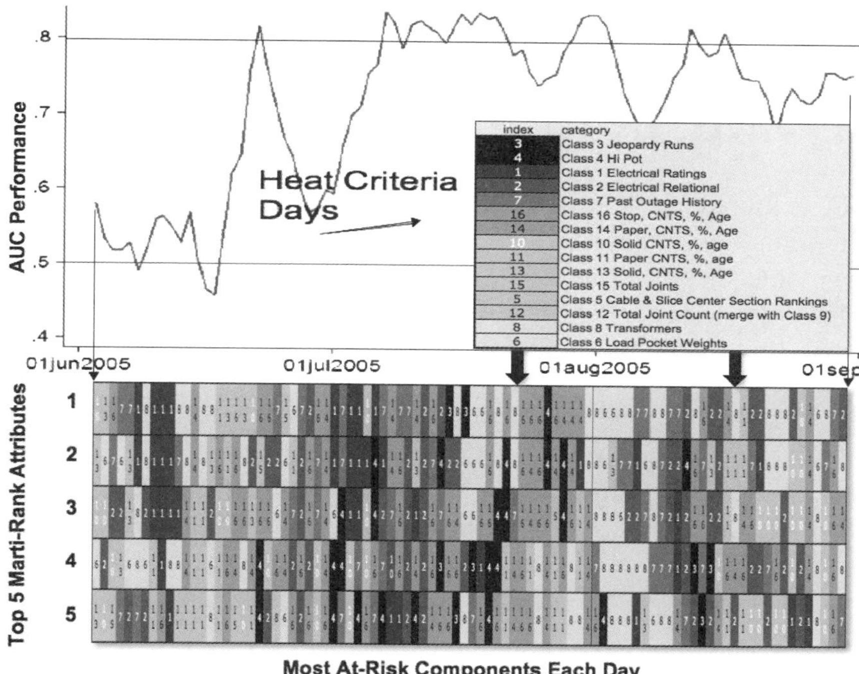

Fig. 9–4. Performance accuracy of the ML ranking system from July through August of 2005 (top). CNTS: counts of the numbers of each attribute class that are present on each feeder. Patterns of component failure predictions can be seen (bottom).

The prediction performance by the ML system also indicates the top-five attribute classes contributing most to susceptibility (fig. 9–4, bottom). In this example, the top-five worst attribute classes change from day to day, as heat waves come and go during the summer. At the beginning of summer, electrical ratings of components and load distribution are most important (dark); then, when the heat begins in earnest, there is an increase in importance of joint and section failures (gray); and finally the transformer class of attributes emerge as important later in the summer (light). This knowledge can lead to insights into what components are most in need of winter replacement beore the next summer's heat arrives.

CONTINGENCY ANALYSIS AND VARIANCE DETECTION

Operational personnel need to know the specific electrical distribution feeders that are most susceptible, as well as most critical to operations, at any given time so that they can take action to prevent outages from happening, if possible. Clearly, resources should be targeted to feeders most at risk of failure and at risk of affecting customers. We are developing a contingency analysis dashboard (DYNACAP) for control-center operators that will guide maintenance crews can be focused on the overloaded components that are most at risk of actually failing. Can the effectiveness of money spent on maintenance and replacement be improved—maybe even doubled—by accurate rankings of risk in the future? These are the questions that are beginning to be answered with DYNACAP (fig. 9–5).

The ML system has shown an unexpected correlation between impending failures and high load pocket weights (the sum of open mains, the low-voltage cable feeders that are out of service, and banks of transformers that are off-line). In a crisis caused by the combination of equipment failures and a severe thunderstorm that quenched the NYC electrical system on a hot week in the summer of 2006, immediate susceptibility increases were evident in response to increasing load pocket weights (fig. 9–6). As feeders became more at risk, they jumped to the top of the ranking list (top) because load pocket weight increased from transformers out of service and low-voltage open mains that were disconnected by blown fuses or secondary cable failures (bottom). The thunderstorm accompanied by torrential rain then quenched this overstressed system, and 10 of the most-at-risk feeders in the neighborhood network failed. Slowly over the next week, repair crews brought the failed feeders back online, and both feeder susceptibility and load pocket weight returned to normal. The susceptibility crisis ended.

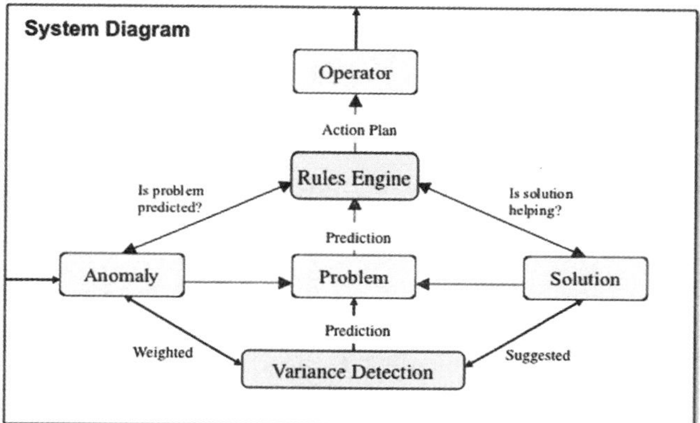

Fig. 9–5. The DYNACAP tool. The ML system, variance detection, and a rules engine are used to discover and push most-at-risk information to the operators with few if any mouse-clicks.

The real-time ML system has been learning the causal mechanisms for feeder failures since it was first turned on in its primitive, "static" form in 2005. Back then, a single martinrank boosting model was used. It was quickly realized that the electrical system changes its causality of failures over the seasons, so a much more sophisticated, 100-model "concept-drift" system was invented and put to first use in 2007.

We have also deployed detection software that measures variance from expected performance of the transformers. Whenever a feeder is taken out of service for any reason, the electric load is transferred to nearby support transformers in order to keep the power on. These transformers feed a common secondary low-voltage grid that powers customer homes and offices. The electric load is measured at all times by the SCADA system for every transformer in NYC, so if the amount of load going to each nearby transformer is out of specification, the control-center operator can be notified. This variance detector has proved useful for identifying previously unknown open mains in the low-voltage grid that could cause a loss of service to customers in the future. These open mains can directly lead to flickering or dimmed lights or even outages. It is believed that several outages have already been prevented through the use of CALM tools.

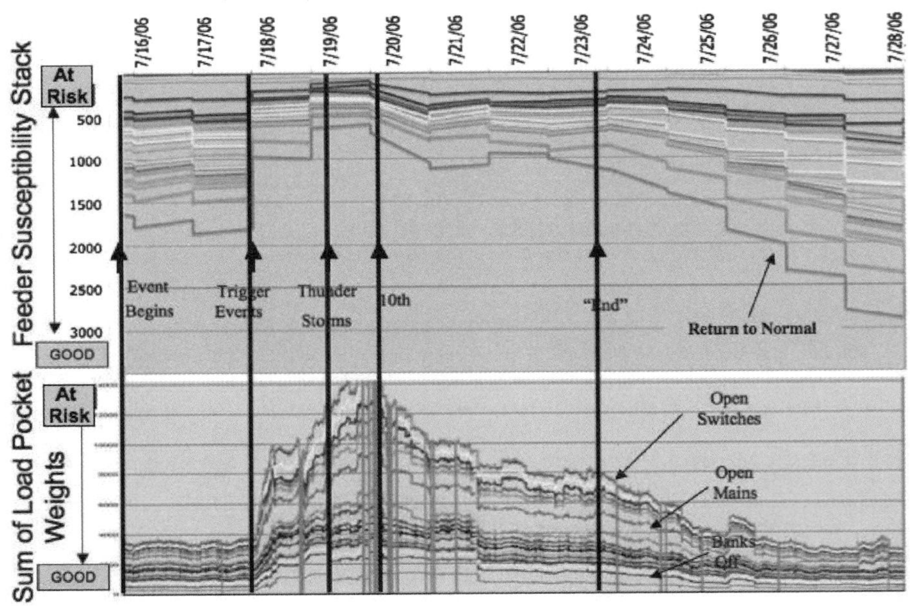

Fig. 9–6. Susceptibility rankings of 22 feeders in the network in crisis, tracked over a week in July of 2006. In the top plot, the rankings of the feeders are stacked on top of each other from day to day in order to show the cumulative risk for all feeders that remained in service in this network during this crisis. In the bottom plot, the sum of transformers and low-voltage open mains that were out on each feeder are stacked.

Load information for each transformer is obtained at roughly 15-minute intervals by the variance detection system. A time-series display of the shifted load results is created, enabling an instantly available view of the state of load transfer after any feeder is taken out of service. A Web page was developed to show the topology of the feeder (what equipment is connected to what and in what order). As shown in the example shown in figure 9–7, the substation is at the top, and electric current flows from manhole to manhole via feeder sections to the transformers at the bottom of the display. If the feeder bifurcates, the rows divide and subdivide into smaller branches of the tree as the current is partitioned among more and more cable tributaries.

Feeder Run from SubStation, to Manholes, to Transformers

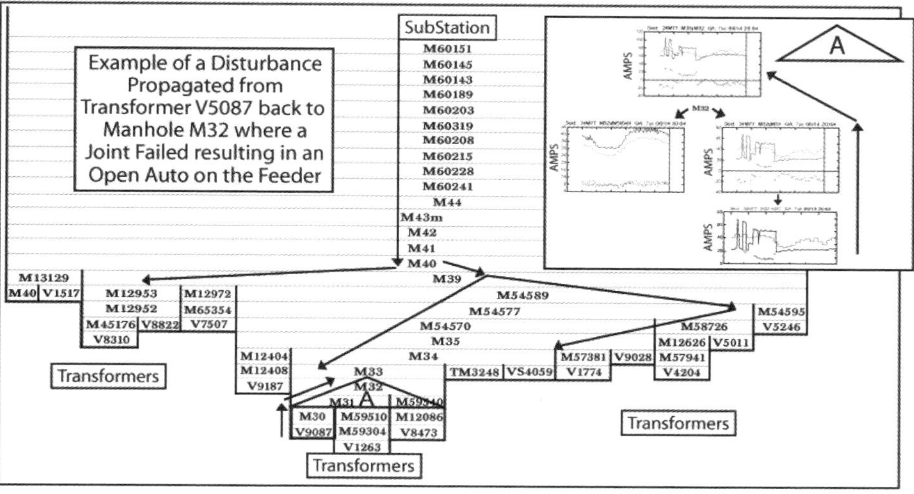

Fig. 9–7. The topology of an electrical distribution feeder—displayed from the substation (top), through manholes (M), to transformers (T and V designations) and the customers (bottom). Clicking on any structure produces a time series of the amperage carried by the feeder over the time interval displayed (see blowup in A). The abrupt end at the vertical line is the time when the feeder went out of service (in this case, a feeder failure or open auto). The smoother background line in each graph is the prediction for electric load on that transformer or manhole for that day. The symbols at the bottom of each plot measure the variance between the actual (jagged) and expected (smoother) loads.

Each structure has a hyperlink to multiple time-series graphs (fig. 9–7). Note that in the example, it is easy to see the massive "chatter" along the M33/M32/M31 manhole path starting approximately 22 hours prior to the failure of this feeder. By following the chatter along the downstream links, operators could trace the problem to a transformer in vault V5087 (the anomaly is visible in fig. 9–7, blowup A). Problems can be seen at the transformer, which produced wild current swings before it actually went dead 14 hours before the feeder failure occurred. The failure was from a blown joint in manhole M32. Manhole M32 felt the chatter because the disturbance was propagated all the way to the substation, and that might have caused the joint to fail. The chatter did not branch to the M59540 structure.

This type of real-time monitoring, when coupled with ML and associated computational algorithms, can result in condition-based maintenance that can help prevent and mitigate unplanned, emergency feeder failure incidents, as well as significantly reduce maintenance expenses. A CALM preventive maintenance program would dispatch a crew to this transformer site immediately upon detection of the variant behavior.

We wish to eliminate "Sherlock Holmes events," or requirements of the operators to perform excessive deductive reasoning to convert clues to discoveries. Just as on a battlefield, better situational awareness of unfolding events that could be indicating additional system failures results. Control-center shift managers have documented that 30 or more mouse-clicks are required to manage such searches within the current IT systems of some major utilities. We also found alarm monitors with up to 36 different alarm categories that are constantly demanding operator attention in utility control-centers. CALM alarm management practice acknowledges that human operators can respond to only five or fewer alarms going off at any one time. There are many examples from other industries that have quantified and documented tools like contingency analysis and variance detection that result in major ROI improvement.

TIME-TO-FAILURE PREDICTIONS

An extension of susceptibility rankings is to predict how much time is remaining before components will fail, termed *survival analysis.* We estimate the *mean time between failures* (MTBF) of feeders by integrating the survival analyses of individual components of feeder sections, joints, and transformers. This provides a prioritization that is useful during the planning of capital reinforcement and replacement programs for components of the aging electrical infrastructures in urban centers throughout the world.

Information about the health of the various components being considered for work is required in order to prioritize what to replace first. We have used the ML system to conduct a survival analysis of cable sections and joints. We have ranked the likely remaining time before failure of all feeder sections and the various configurations of joints (differing by composition, design, equipment manufacturer, and age) in the NYC electric grid. These predictions are then combined along the length of each feeder to produce a capital asset prioritization tool CAPT. CAPT also incorporates electric load information on sections. The most-at-risk sections and joints should be the first to be changed out of the system, all other considerations being equal.

We trained on outages from 2001 through 2005 and tested whether our susceptibility-ranking system was predicting the most-at-risk components that would fail by holding out one-third of the outage data sets to verify that ML rankings were working. We then conducted two different blind tests on each component data set during the summers of 2006 and 2007 (fig. 9–8). When each new feeder failure occurs involving a section, the ranking of the section that failed is put into its appropriate ranking bucket. The system is performing adequately because the histograms in figure 9–8 are skewed to the most-at-risk (left) side, meaning that they have successfully predicted more than they got wrong. An equal (horizontally flat) distribution from most at risk to best would represent random predictions. For example, 55% of cable sections that failed in the blind tests were from our most-at-risk 20% of the bucket of all 150,000-plus sections citywide in the blind tests conducted in July and August of 2007, respectively (fig. 9–8). The ML analysis was performing well enough to be used in section and joint replacement decisions beginning in the fall of 2007.

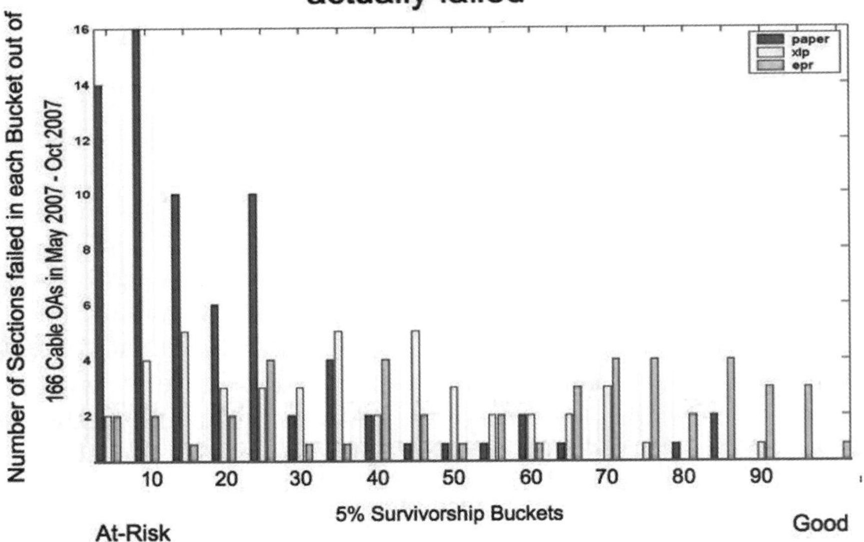

Fig. 9–8. Blind test conducted on component data sets during the summer of 2007. The ML rankings are blind tested against new cable section failures that occur after the rankings were completed.

The CAPT locates each section and transformer in latitude and longitude, transfers the rankings to Google Earth, and "paints" onto feeder locations the type and vulnerability of each cable section. Rankings then provide a citywide view of which cable sections most need to be replaced quickly. For example, vulnerable paper-insulated lead-covered (PILC) sections near the substation in figure 9–9 were identified and replaced. After the replacement, performance of the feeders during the summer of 2007 was much better (see fig. 9–9).

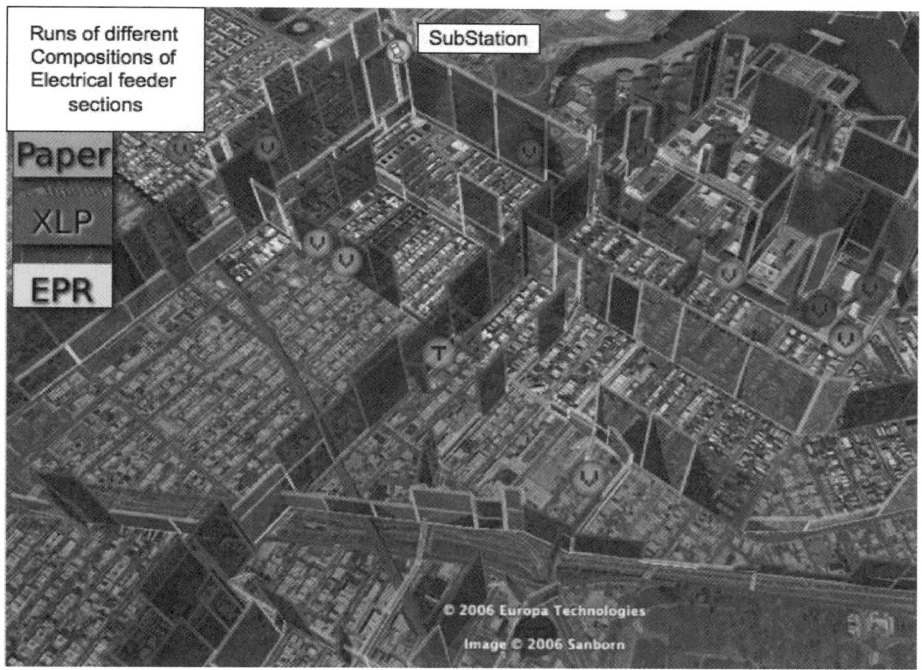

Fig. 9–9. The composition of feeder sections in a network of NYC, represented on Google Earth by shading. The height of each section represents its ranking, with those most likely to fail the soonest shown with more height. The dark gray circles are vulnerable transformers, and the light gray are good transformers. XLP and EPR are newer types of cables used for electric distribution feeders that run under the streets of NYC.

The CAPT permits users to examine the existing state of a network or feeder and visualize useful attributes of the system, ranging from feeder susceptibility ranking to cable and joint rankings and predicted future loading (fig. 9–10). The user can operate in three modes: reinforcement, PILC replacement, and reliability. In the latter two, there are many choices that could potentially satisfy policy requirements.

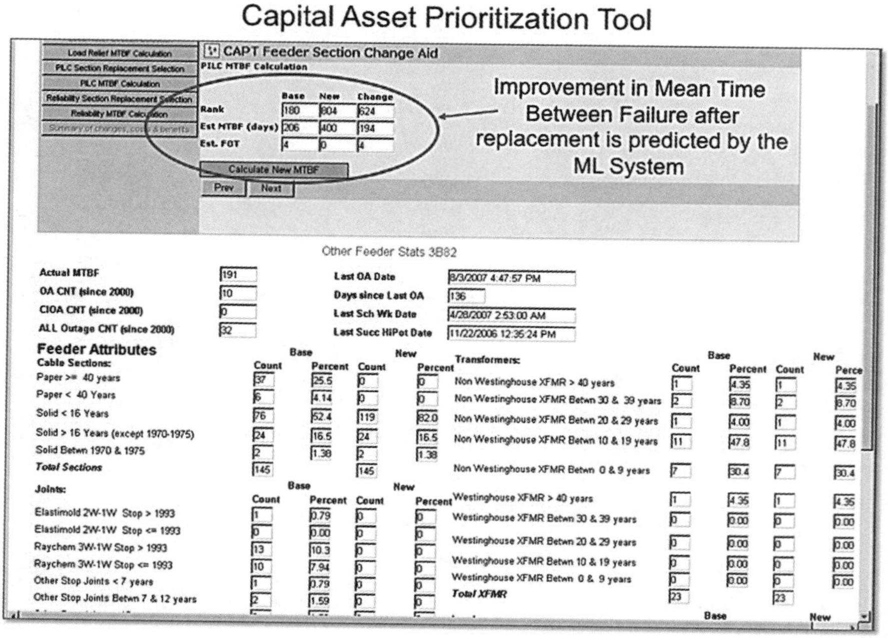

Capital Asset Prioritization Tool

Improvement in Mean Time Between Failure after replacement is predicted by the ML System

Fig. 9–10. Screenshot of the CAPT. The rankings of feeder sections, joints, and transformers can be used to evaluate costs versus predicted MTBF benefits in order to plan the right work during the PILC section and joint replacement programs.

Our survival analysis–based MTBF technique estimates the improvement in feeder performance that can be expected after various strategies for replacement of old PILC cable sections and their joints are carried out on feeders. The CAPT will in the future track the actual performance to test if the benefits were worth the costs using classic CALM techniques. The CAPT system feedback will be in terms of both reliability improvements using the MTBF metric and cost evaluations of alternative actions. This platform provides the basis for reliability-cost maintenance programs being put in place by Con Edison.

By use of another novel ML technique, a new algorithm called *support vector censored regression* (SVCR) was invented to estimate MTBF.[1] The likely MTBF of all sections and joints in Queens was calculated in order to predict the 33 most-at-risk feeders (fig. 9–11). A blind test validation of the MTBF of the most-at-risk ranked components will be used to validate that the CAPT discovers the best strategy for replacing poorly ranked cable sections and joints on such feeders in the future.

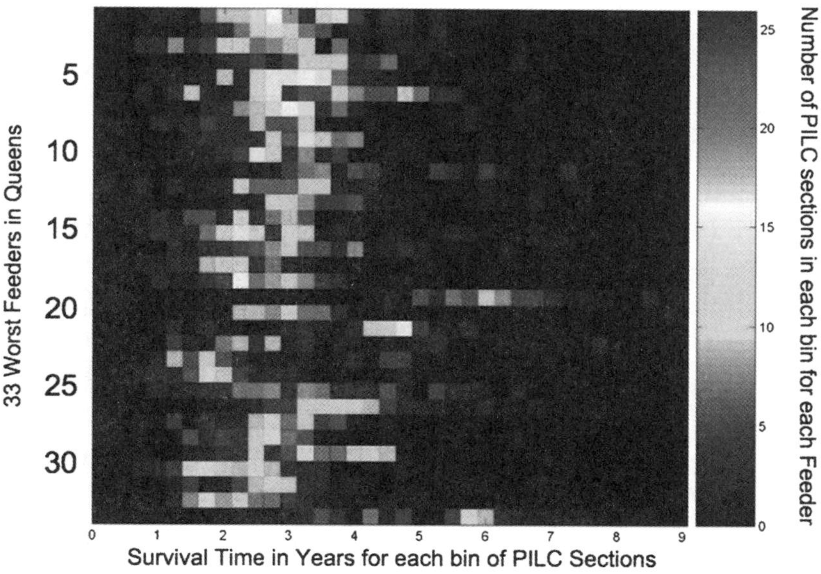

Fig. 9–11. Survival analysis used to predict the 33 feeders with the most-at-risk sections in Queens. The number of sections in each feeder is indicated by the color in each bin.

BACKBONING FEEDERS

We then tested the hypothesis that eliminating almost all of the PILC (paper) sections and their joints (vulnerable because they are mechanically complicated) on specific backbone feeders would lower the frequency of future emergency outages by increasing MTBF, and thus enhance electrical network reliability. Backboning is a key strategy for strengthening neighborhood network reliability. We needed to develop a methodology to look at existing performance data to answer the question: what percentage of the PILC sections and their accompanying joints would have to be replaced to gain the dramatically improved performance benefits that would justify the costs of backboning?

First, we compared the outage history of all network feeders after they were put into buckets according to their percentage of stop joints (fig. 9–12). We are looking for a possible "purity minimum" classification in the existing feeder stop joint counts that might correlate with the right percentage of stop joints and their PILC cable sections that would qualify them for backbone feeder status (i.e., lowered feeder failure likelihood in the future).

In figure 9–12, 35% of the feeder population with only 0–5% PILC joints (diamonds) had zero failures, whereas 20% had only one failure and 15% had two failures. This was outstanding performance when compared to any of the other feeder groups that are more densely populated with stop joints (fig. 9–12). Backboned feeders with a population of 0–5% PILC joint count have a statistically better performance than those with more PILC joints, even for the 5%–10% group. We now know how extensively a feeder must be purged of PILC joints (eliminating such joints will also eliminate PILC sections, of course) to qualify for backbone feeder status.

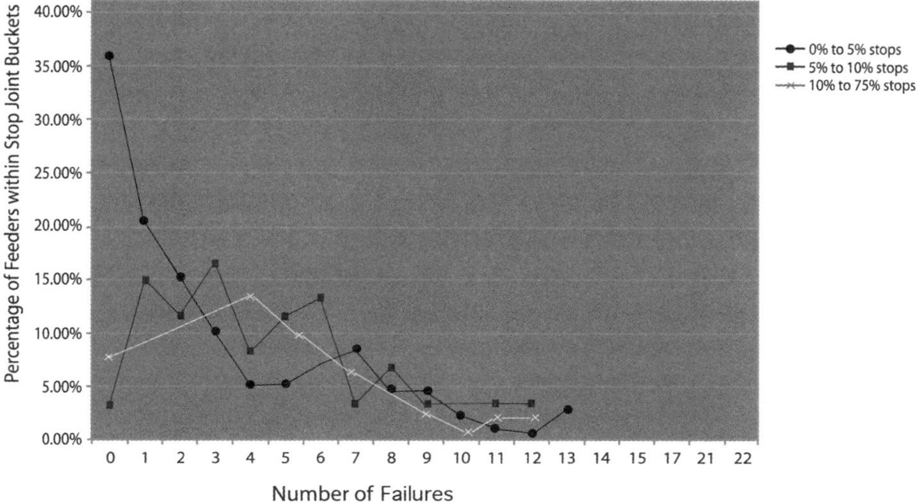

Fig. 9–12. Histogram of the number of times feeders with various percentages of PILC stop joints (collected into buckets) failed for all network feeders since 2001. The population with less than 5% stop joints performed decidedly better than all the other groups.

We then used a well-known statistical technique called *propensity score* to match pure backbone feeders with a control group of impure ones. We define *impure* as those feeders containing a mix of different types of sections (both PILC and newer compositions) that otherwise have similar failure statistics, lengths, daily load stresses, and electrical rating except that they have relatively more PILC cable sections and their joints. In other words, we chose a set of impure feeders that, although of mixed compositions, are otherwise similar to the pure feeders containing less than 5% stop joints and, therefore, few PILC feeder sections. The attributes used for comparison were load pocket weight, shifted load support, and total number of joints. These three attributes were chosen because they were independent of cable type, and they correlated with failures from our analyses of the summer of 2005.

We chose the set of impure and pure feeders so that similar propensity scores would assure similar distributions of other attributes. We then tested the propensity score matching by comparing the distributions of attributes in the pure feeders and the matched set of impure feeders. For all three attributes, the distributions were very similar for the pure and matched impure feeders.

We then demonstrated the utility of propensity scores for developing a twin-study methodology for addressing existing or proposed capital improvement strategies such as backbone feeder purification. However, our previous study confirmed only part of the operational strategy. The CAPT quantifies other operational requirements such as the identity of the exact feeders to purify and the cost-benefit analysis from the effort required to create backbone feeders. The O&M cost of the quantity of materials and crew time required to create a backbone feeder is evaluated against the savings from an increased MTBF and, therefore, the risk of fewer failures. The reduced business risk is computed from the lowering of failure events within networks, particularly in succeeding summer months.

CLOSING THE FEEDBACK LOOP

CALM provides the mechanism to close the feedback loop on whether the work prioritized and executed via the CAPT was indeed effective. Decision aids such as the CAPT consistently cost in the millions to develop and return performance that saves future outages that cost in the tens to hundreds of millions, for an ROI of 10:1 to 100:1. However, the predictive accuracy of the rankings must be constantly monitored using blind tests. As each component fails in the future, its ranking gives a running tally of how well the survival analysis is performing. The rankings of the PILC sections that were actually replaced upon the recommendation of the CAPT have been tracked as an optimal strategy was developed and tested first in Manhattan in 2006 and then in Brooklyn in 2007 (fig. 9–13). The most-at-risk PILC sections were removed, and now the CALM system is tracking the MTBF of the repaired feeders, which should increase with time now that better components are in the feeders.

ML Ranking of PILC Cable sections that were replaced

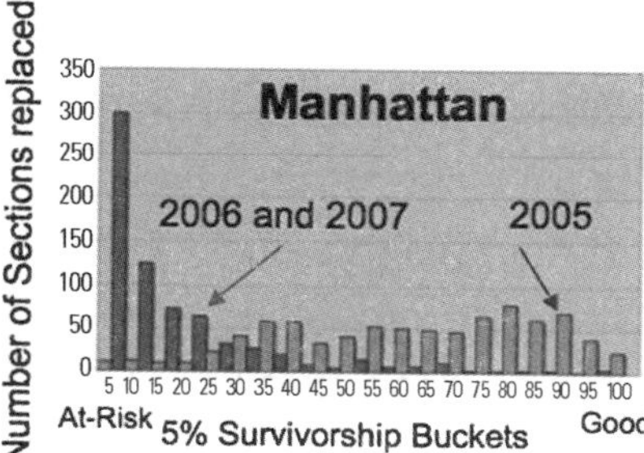

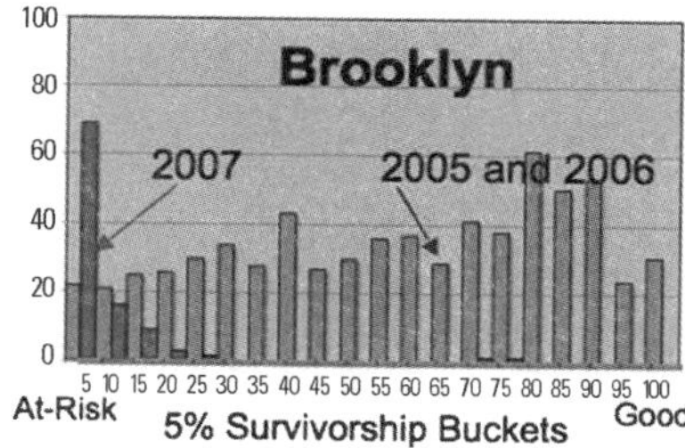

For maximum effectiveness cutting future failures, replace sections ranked in worst Survivorship Buckets

Fig. 9–13. ML ranking of PILC cable sections that were replaced. For maximum effectiveness in increasing the MTBF of future failures on feeders, the CAPT recommends replacing PILC sections ranked most at risk by the survival analysis.

We analyzed all outages on the same feeder and on nearby feeders within each network and found that the failure events, whether electrical or mechanical, cause more failures on the same feeder and on nearby feeders within the network than a random failure model would support. We also found that scheduled work causes more failures on that feeder than the random model would predict, but not on other feeders within that network.

PLANT MODEL FOR NYC

A new PM based on DEW has been constructed for several neighborhood networks in NYC, and it is forming the foundation for an ISM. Con Edison has a legacy power-flow model called *power voltage load (PVL)*, but it is not suitable for the PM. Both the PM and PVL are successfully running power-flow algorithms for base load and next-contingency analysis cases of what might happen next on the grid. PVL is being used to validate that the electrical loads and component performances are accurately reflected throughout the new PM.

The past deployment of power-flow modeling software tools could not provide good estimates of feeder power flows under base load case conditions in the complex combination of overhead and underground, 33-kilovolt (kV) to 4-kV distribution system in NYC. Convergence of these models to solve for feeder power flows under first or second feeder outage contingencies was not easy in the past, resulting in significant risk of not fully understanding the dynamics of various power systems under emergency conditions. The new PM allows accurate power-flow analysis from the transmission grid to all customers on each network for the first time.

The PM extends from the transmission system, to the subtransmission system, to transmission substations, to area substations, down to all distribution substations, high and low outage feeders, transformers, and all the way to customer loads. Figure 9–14 shows an example of the PM that has been successfully built and tested for several networks in NYC. The PM is becoming the core of the ISM for power utilities like Con Edison.

The PM can be used to accurately and reliably investigate the distribution of loads and to identify potential risks in the present design basis of the electric grid. Appropriate reinforcement is now being planned in the winter for operational improvement that must be in place by each next summer.

Fig. 9–14. PM of several NYC electric distribution networks. Light represents 4 kV, dark 13 kV and 33 kV feeders.

A prototype model of a portion of the electric distribution system has been integrated with real-time measurements from the SCADA system, providing the capability to display the results of power-flow analysis such as voltage and current-flow estimates on the mapping system. Inaccuracies will become evident as operators and engineers use the model on a daily basis to run the electric system. Discrepancies will be resolved as the system learns which improvements will result in best operations. Reduced business risk will be the result.

System operators need accurate estimates of voltages and currents throughout the distribution system, down to the distribution transformer level in order to make optimal decisions. The ISM is the foundation for operational innovation because it enables operator aids, such as power-flow management, switching orders, capacitor bank balancing, and reconfiguration for restoration analyses.

The PM (DEW) was used as the calculation engine to perform real-time power-flow analysis at every point on the system at Con Edison–owned Orange and Rockland Utility, in suburban NYC.. The weather-dependent load research statistics, kilowatt-hour measurements, and large customer kilowatt load measurements that are interfaced to the DEW circuit model were used to create a virtual SCADA system. Given existing weather conditions, historical load measurements and load research statistics were used to estimate loading for each customer in real time. The virtual SCADA system then scaled the estimated loads so that the real-time power-flow calculations matched the substation or start-of-circuit measurements. The mapping system was then used to display estimated voltages and currents throughout the circuits in real time.

The PM was used for both the integration between DEW and the SCADA system, as well as between DEW and the mapping system (fig. 9–15). DEW provided voltage and current estimates to the mapping system for display, as well as alarms and switching plans. Real-time power-flow calculations in DEW, using the virtual SCADA system load estimates, were run on 17 feeder circuits. Voltage and current-flow calculations were displayed on operator-selected distribution transformers and other circuit elements in the mapping system, enabling monitoring and analysis of circuit performance that was never before available to operators and planners.

A Web-based weather-forecasting service was integrated into the PM by EDD to provide 24/48-hour-ahead weather conditions for use by operations to prepare for emergency deployments of crews and equipment. Electrical damage and customer outages due to severe weather are costly in terms of dollars and public inconvenience. An ISM that responds to storm conditions can be used to predict outages and system reliability. The PM made operations more efficient, with the net effect of lower operational costs and shorter customer outages. EDD, Inc. (creator of the DEW PM), attached storm models to predict outages due to approaching storms and to calculate realistic failure rates and repair times in advance.

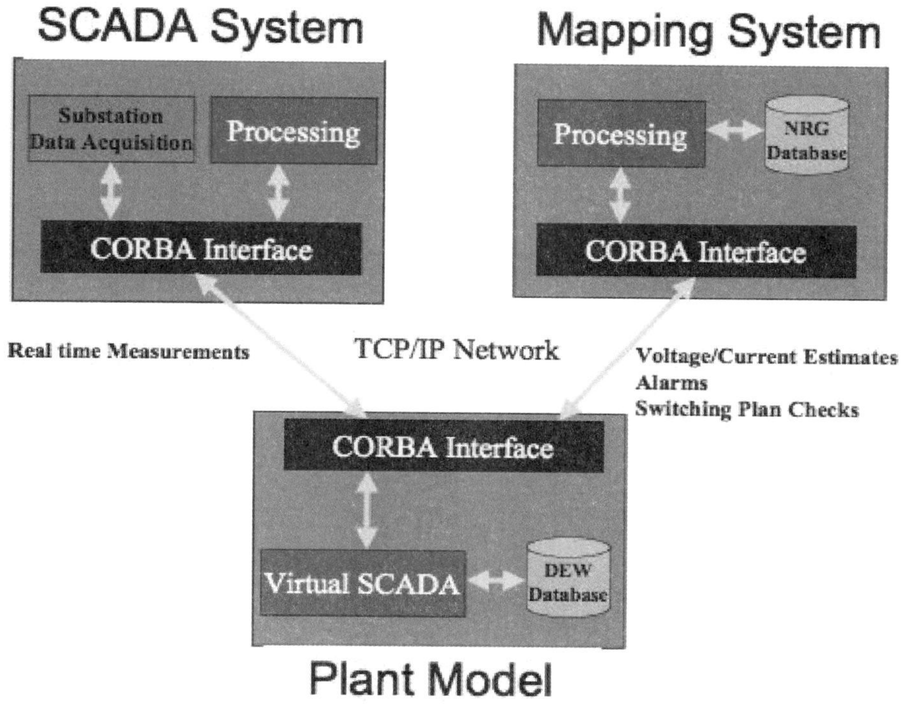

Fig. 9–15. The PM controlling the interactions of the virtual SCADA system with the real SCADA and the mapping systems

The approach was to use historical storm and lightning data to correlate this with historical outage data. As part of this work, storms have been classified into different types based upon temperature, wind speeds, and lightning activity. Specific types of storms analyzed were high temperature; high temperature with strong winds; and high temperature with strong winds and lightning activity; and so forth. Altogether, 15 different categories of storms were considered. Outage data since 1995 were used in the analysis.

Expected outages per storm hour for high-temperature storms and high-temperature storms with strong winds are shown in figure 9–16. An interesting result was discovered: outages start to level off after 20 hours into such a storm. Subsequent analyses found this to be true across all storm types analyzed.

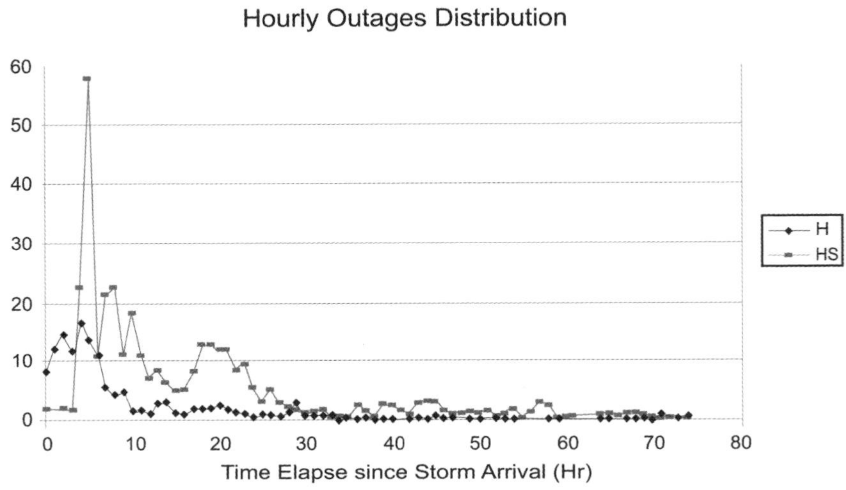

Fig. 9–16. The number of outages per hour, for high-temperature storms (H) and high-temperature storms with strong winds (HS)

This is even true for storms that last for several days for both high-temperature storms and high-temperature storms with strong winds (fig. 9–16). Figure 9–17 shows average downtimes in minutes for six of the storm types analyzed.

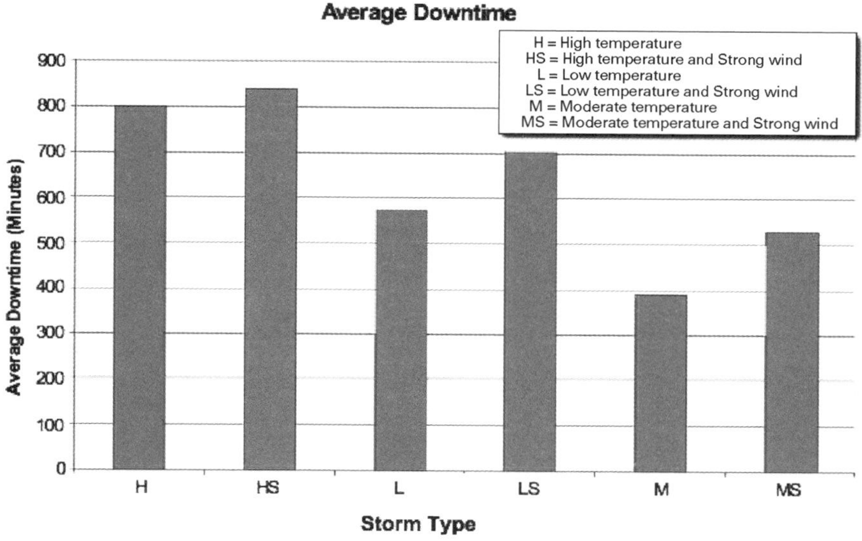

Fig. 9–17. Average downtime of customers without electricity per storm, for storms of each type

In storms involving lightning activity, corridors around transmission lines were monitored for lightning hits, and the lightning densities within varying corridor widths were determined. Analysis to date indicates that 200 feet is a good choice for the width of the most likely at-risk corridor to be considered around transmission lines. The lightning density within transmission corridors is always higher than the density over the entire storm area. This seems to indicate that the transmission lines attract lightning. Another interesting result is that the majority of strokes that result in outages are more than 25 feet from the line. Finally, strokes with intensities in the range of 10–30 kiloamperes cause the majority of outages.

EDD's analysis of the lightning data shows that outages due to lightning are almost a linear function of the lightning density within transmission corridors. A plot of the lightning flash density versus expected number of outages is shown in figure 9–18. Note that there are not many lightning-intense storms in the sample, and these particular results are not statistically robust but require more lightning-storm samples for a higher degree of statistical confidence.

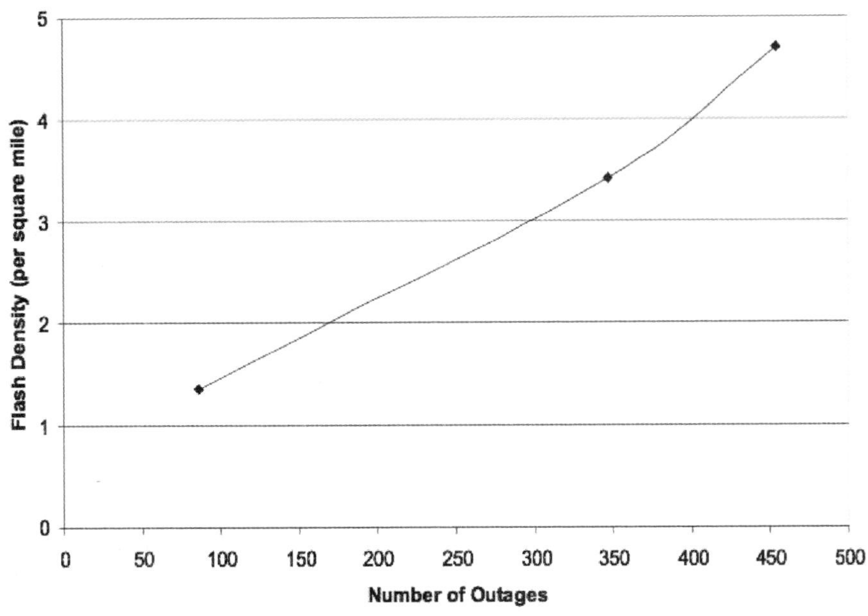

Fig. 9–18. Relation between lightning and power outages during electrical storms. Flash density in 200-foot corridor of transmission lines is plotted versus number of power outages.

The ISM can be used to predict outages that are going to occur due to approaching and/or existing severe storms. Real-time data to implement this functionality are available from two different interfaces from WeatherBug—one for normal weather variables and the other for lightning activity. If the storms involve lightning activity, then the real-time lightning data feed can be used to calculate lightning densities within transmission corridors. Future innovations will focus on evaluating equipment failure rates and repair times based on the real-time conditions of weather on the system, as well as crew availability.

Finally, Con Edison is also testing IBM's Deep Thunder weather prediction software for storm management, augmented with additional weather stations that Columbia University has set up for their use at key substations in the NYC area. Deep Thunder's resolution, down to a resolution of 1 kilometer (km), is better matched to storm damage prediction needs than the model predictions from the National Weather Service, which has only 12 km resolution. It is expected that real-time modeling will be part of the real-time ISM analysis of electric system reliability in the near future. Worker dispatch will benefit from better lightning and weather forecasting that are integrated into the operational ISM.

NOTES

[1] Shivaswamy, P. K., W. Chu, and M. Jansche. 2007. A support vector approach to censored targets. Presented at the IEEE International Conference on Data Mining (ICDM-07), Omaha, NE.

[2] Rosenbaum, P. R., and D. B. Rubin. 1983. The central role of the propensity score in observational studies for causal effects. *Biometrika.* 70: 41–55; D'Agostino, R. B., Jr. 1988. Tutorial in biostatistics: Propensity score methods for bias reduction in the comparison of a treatment to a non-randomized control group. *Statistical Medicine.* 17: 2265–2281.

[3] Zhu, D., D. Cheng, R. P. Broadwater, and C. Scirbona. 2007. Storm modeling for prediction of power distribution system outages. *Electric Power Systems Research.* 77: 973–979.

10
GROWTH

CALM uses software for the optimization of profitability under uncertainty. In this chapter, we show through example that decisions on growth of a firm should consider all aspects of the enterprise, including future markets and regulation, alignment of personnel incentives with corporate objectives, employee wellness and happiness, and customer satisfaction. Otherwise, major disconnects can happen that cost even a lean company dearly. For example, we have seen that Boeing is one of the stellar lean companies, but it suffered stockholder and customer ill will when it underestimated complexity issues with new, lean, plug-and-play manufacturing designs of its 787 Dreamliner. Also, Toyota was struggling with quality control, of all things, for its automobile manufacturing as it grew into the world's largest seller of cars. Constant vigilance, monitoring, evaluation, and reevaluation of performance are required by CALM. We start this discussion with proposed decision-making for new asset investments and follow with examples of often-overlooked opportunities and impediments to growth in enterprise value.

ASSET INVESTMENTS

In general, businesses decide how to allocate limited resources to grow enterprise value across a portfolio of potential investments that provide real options for growth. Portfolio analysis for the financial markets has matured over the past 30 years so that investors can now efficiently manage shares in a continuously changing portfolio of companies. Portfolio managment calculates risk versus reward so that each investor can satisfy their own personal efficient frontier. Advancements in the modeling of risk, such as the *capital asset pricing model* (CAPM) and the *distance-to-default* metric (probability of default is determined through financial statements and stock prices), have proved to be effective evaluators of the risk (volatility) versus value of investments for individuals. Wall Street has mastered value at risk (VaR) computations to determine, with statistical confidence over a certain time frame, their risk profile of investments based on historical data. However, these models and metrics by themselves are insufficient for determining strategic investment decisions in hard assets when looking from within an energy company.

Investment considerations within a business need to be based on underlying competitive advantages such as providing differentiated levels of service or added product value through better internal business capabilities that competitors do not possess. A company is supposed to have certain core competencies and take on those risks that it is most capable of managing in order to increase profitability. Thus, portfolio analysis from within the firm takes on a different meaning when developing strategic capital allocation plans for growth.

Consider long-life assets like oil platforms, gas pipelines, electric distribution systems, or electric generation. The company's success is tied to ongoing investments in sustaining the reliability and profitability of these assets while mitigating risks of failure, some of which can be catastrophic. Investments in such long-life assets are inherently risky because the firm essentially takes a "long" time duration position in commodity supply, while being susceptible to future demands of customers who take a "short" time duration position. Customers buy the commodity or service depending on what is happening to them on a daily to monthly basis. They use contracts and derivatives to hedge some fraction of their supply needs for one to two years out, if they are smart.

To make portfolio management even more challenging in the energy business, markets and regulators of commodity-dependent assets are location specific. Therefore, competitor actions and government regulation changes may significantly affect the company's assets in one location but not another. Identical assets in different geographical locations can have completely different risk/reward profiles.

From strategic management (SM) perspectives, existing infrastructure-intensive businesses, for the most part, have their strategies "poured in concrete." The investments in assets made over the past few decades and their specific locations have created a rock-hard foundation for much of an energy company's strategy. This leaves only the opportunity to place incremental bets using limited investment dollars to attempt to "move the needle" of value creation by entering into a few new investment opportunities while also taking care of existing investments. These incremental investments are essentially the only path for sustaining the growth of an infrastructure-intensive firm if it does not resort to considerable divesture of existing assets. For this reason, many infrastructure-intensive firms look to their business units to provide the bulk of the strategic planning. The business units, focused by definition on growth of their existing business, create yearly business plans for new capital allocation primarily by tweaking last year's budget on the basis of estimates of how their local world is forecast to change.

Business units serve up investment opportunities that are (correctly) focused heavily on who they are already, whether it is a power generating business or an oil exploration business. The issue at hand is how an infrastructure-based business can develop new opportunities for growth while minimizing risks? One of the CALM keys to SM is to view the firm as a single compound option using a portfolio of real options to augment traditional portfolio analysis.

Another challenge to SM for energy companies is the "herding the cats" problem with semiautonomous business units. These multiple business units should be providing coherence and synergies that create interrelational value for the whole firm, but they often are incentivized to think only of their own interests. SM within energy companies by its nature requires a centralized decision point for portfolio analysis, beginning with consideration of the "growing from what you know" principle of business development and ending with how to leverage value from multi-business entities. The imperatives of growth require a committed effort at SM, which includes detailed analytics of the state of the company's worth, strategic options, interrelational effects, and performance using a real-options framework that can also validate competitive advantages the company perceives it may have.

Within SM, portfolio and financial risk should be coupled with measured risk from interactions among its operations and assets. Diversification for the sake of mitigation of stochastic variability in projects is a waste because it could have been done by the company's investors. Instead, investing with clear understanding of operating options available to the firm can create positive quantitative interactions with related projects that are of high value. Especially in the oil and gas industry, the cash flow from one project can be positively or negatively affected by performance of other similar geological or geographical projects and therefore needs to be taken into consideration.

However, for all energy industry infrastructure companies, risk management and the management of real options are not just overlays to consider in making strategic decisions on investments, but critical considerations. Without them future growth is often preempted in both existing businesses and in diversification strategies aimed at leveraging its competencies. For instance, high systemic risks in a volatile market can be profitably managed through higher investments in real options, whereas in less volatile markets, a company may accommodate a leveraged asset investment when interproject support is possible. The reasoning for heightened measurement, management and control of risk, using options is that the company can be sitting on investment time bombs, to be triggered by uncertain events that can rapidly convert previously stable long-life assets into liabilities.

Real options take on significant value in this type of uncertain world. They make the enterprise agile via the flexibile value of real options. Importantly, they can add some insulation against the inevitable "black swans"—improbable events that cannot be anticipated or modeled but that have high negative impact. A *black swan* is the term in the financial community for a train wreck that could never have been seen coming.

The CALM approach would be to utilize ISM simulations and scenario analysis to generate insights into how and where to find high-impact improvements in real options for future growth. The intent would be to use an ISM of the business to efficiently determine realistic strategic, operational, and financial risks and opportunities faced by a company through simulation. Managing these real options consists of pricing them repeatedly over time and executing on real options based on an optimal exercise policy that makes sense within the firm's strategic vision.

Four points are critical to successful use of portfolio management tools to ensure that misalignments of investments in such real assets do not happen:

1. *Know why you are using portfolio management.* Is this being used for SM or for optimization of a previously defined level of expenditures to replace components or assets within a business? Identify and measure underlying risks; quantify volatility; test feasibility of strategies; identify both significant and detrimental projects; define unique value of investments; evaluate investment options; and continuously monitor business performance.

2. *Know who are the end-users.* Operations portfolio management relies on extensive input and interaction with engineers, field workers, and planners. However, the end-user is the decision-maker, who is responsible for managing overall business performance.
3. *Know critical success factors for implementing portfolio management so that the performance of the company can be continuously tracked.* Establish a clear portfolio process; set commonly understood, clear strategies and goals that are known and understood by all, both internal to the company (employees) and externally (shareholders and customers); and develop strong relationships among key personnel (the soft side).
4. *Always aim for high-quality decision framing.* Clearly define the problems to be solved and the decision criteria to be used to select the solutions; generate plausible real options; evaluate the options relative to the decision criteria; and make the decisions only when the options present a viable path forward. Be brutally empirical.

Better decisions that improve the probability of consistently meeting strategic goals will be the likely result, whether they are operational, financial, engineering, and/or health, safety, and environment (HSE) related.

OPPORTUNITIES AND IMPEDIMENTS

Given an uncertain world, energy companies plan investments with an eye for where markets will be in coming years, and they are constantly vigilant of assets turning into liabilities or becoming obsolete. With detailed measurement and analysis of the present state of the market and of the true state of the company, scenarios of future investments within a well-defined set of opportunities can be evaluated using portfolio management. In the *gas-to-electricity* example in the next section, we show how methodologies including scenario and real-options analyses are used to demonstrate the value of the flexibility that CALM brings to strategic and operational decision-making in times of disruptive transitions in energy markets. In the example, analysis suggests that additional enterprise value can be created through the addition of new environmental innovations, such as adding a gas-to-electricity option for offshore production hubs centered around floating generation, sequestration, and off-loading (FGSO) vessels and undersea high-voltage direct-current (HVDC) power cables to shore. We show that these innovations would not be economically viable except for the crisis of global climate change and the coming global conversion to an electric economy that the Earth is facing.[1]

GAS-TO-ELECTRICITY

The evaluation of future investment options can be based on potential changes in exogenous forces (e.g., supply/demand, geopolitics, and climate change). Scenario and real-options analyses can provide rich alternative investment strategies to consider in today's volatile business world.

The dynamics of geopolitical uncertainty, always a risk in the oil and gas industry and a major source of black swans, are compounded by uncertainties arising from potential consequences of global climate change, hydrocarbon supply/demand shortages, and always-increasing demand for electricity as the cleanest environmental solution. We have evaluated scenarios concerning how to maintain profitability in ultra-deepwater oil and gas production with the world moving toward a green, electric economy, in which there are significant new penalties for carbon emissions and most end-use of all forms of energy is via electricity delivery. In such scenarios, a modern oil and gas company must consider its real options for remaining profitable in the face of potential new carbon taxes, carbon sequestration requirements, and competition from coal gasification, nuclear, solar, wind, tidal, and even space-based, solar power generation.

Given the uncertainties inherent in future geopolitics, technology development, and international competition, analyses show that making significant investments purely on the basis of the existing paradigm of oil and gas market dynamics can easily result in lost opportunity at best—or bankruptcy at worst—in the future. An alternative approach is to use the techniques of scenario and real-options analyses to create alternative investment options to capture out-of-the-box opportunities without making significant commitments of capital investment until absolutely necessary.

Real options in the offshore

Maintenance of a real option to produce gas-to-electricity offshore provides a compelling argument to make potential investments into a new generation of production technologies—namely, floating generation, sequestration, and off-loading (FGSO) vessels (fig. 10–1). Our analysis shows that the opportunity is at least as large as that realized in the past 15 years from the invention and widespread deployment of floating production, storage, and off-loading (FPSO) vessels.

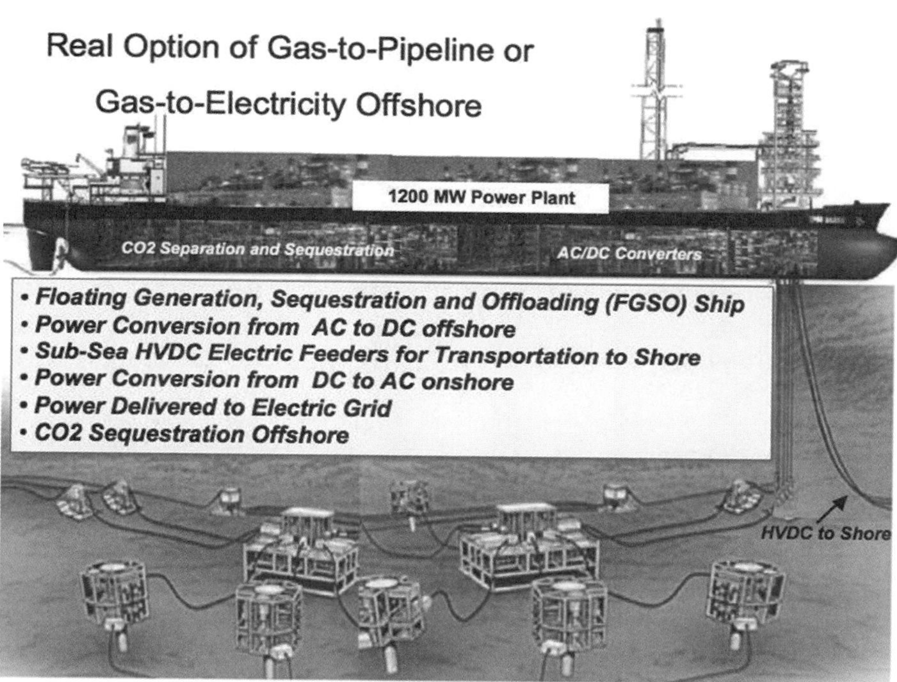

Fig. 10–1. FGSO ship concept, for maintaining the real options of generating electricity from produced natural gas by sending it and/or gas to shore to sell into different markets

Commitment of capital in any resource-constrained business such as the energy industry is similar to storage of products in a warehouse without immediate increases in enterprise value. There is significant risk that the products stored will lose value or become obsolete (e.g., U.S. automobile inventories of gas-guzzling sport utility vehicles). Delaying capital commitments as long as possible as dynamic forces change, while anchoring in the real options to enable entry into emerging markets at a later date, is good strategy (fig. 10–2). Several possible scenarios for future success can be kept open with real options so that dynamic market forces and changes in strengths, weaknesses, opportunities, and threats (SWOT) to the company can be reacted to accordingly.

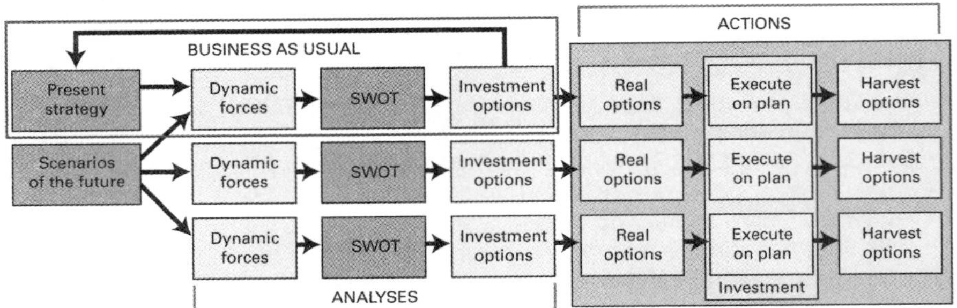

Fig. 10–2. SWOT analysis of keeping multiple real options available to hedge against an uncertain future, compared to the business-as-usual approach arising from NPV evaluations

Once capital investments have been made, reinforcement learning (RL) controllers can be used in the future to provide the mechanism for computing the real option value *in real time* of added capabilities that arise from swapping delivery of variable product mixes to multiple, asynchronous markets. Price signals that vary by geography, as found in the energy markets for natural gas and electricity in the southern United States, qualify for such real-options considerations for ultra-deepwater production from the Gulf of Mexico. To take full advantage of dynamic differences in commodity prices, producers in the Gulf of Mexico would need to add additional real options of selling electricity generated offshore from produced gas, paired with potential trading in CO_2 credits from sequestration right at the source of the production. If price and cost signals are both received in real time, and if sufficiently diverse product and sequestration options exist, then the RL controllers could select most profitable mixes to produce the added value required to justify the large additional capital investment.

Such smart RL controllers with gas or electricity options are especially useful when either positive or negative *spark spreads* exist. A spark spread is the difference in price between gas and electricity in a given market at any given time. In 2005–2007, regional spark-spread premiums of ±$20 per megawatt-hour equivalent between the electricity and natural gas prices were common. These variations were caused by electricity congestion in the summers and gas shortages in the winters. It is possible to construct real options on spark spreads.

The RL controllers will incorporate dynamic real options to compute optimal decision paths for real-time control that are always in the money. When controlling ultra-deepwater production, the RL controllers simultaneously balance price signals against production costs to distribute gas versus electricity into different regional markets in real time (fig. 10–3).

The value of investment in real options comes not from the addition of new kinds of power delivery systems to the market mix, but from the modest initial investments needed to enable the possible selection of options such as FGSO hubs at a later date. The timing and volume of gas and oil streams piped to shore are the only current options available to ultra-deepwater operators. What would it take to generate such additional options of selling electricity generated by burning gas at new offshore power hubs, while sequestering CO_2 back into reservoirs to enhance production and capture possible future carbon credits?

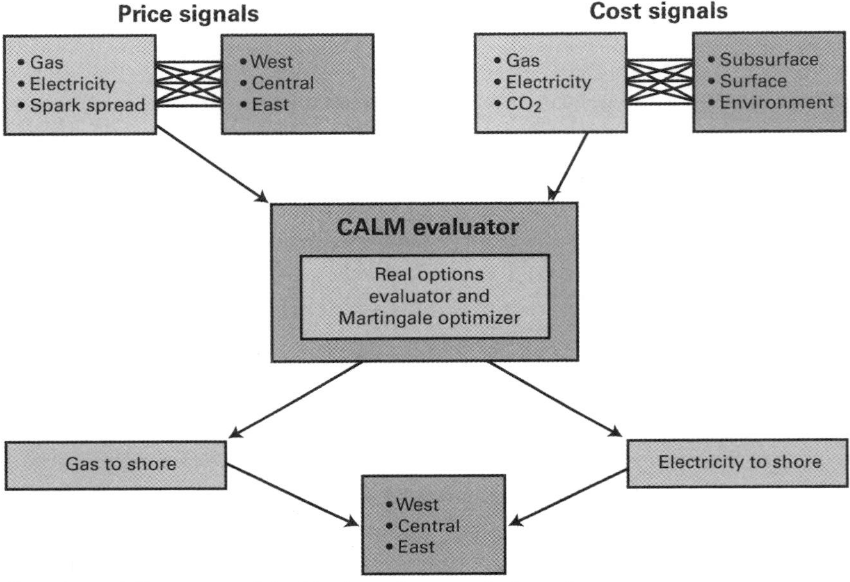

Fig. 10–3. The RL controller that combines real option evaluation and control decisions in the same algorithm to optimize profit from variable price and cost signals

A similar paradigm change in offshore production methodology has been executed by the industry in the recent past. When the need for more diverse production and storage facilities arose in ultra-deepwaters of Brazil and West Africa, FPSO vessels were invented. Today, more than 100 FPSO ships have been built to exploit this need for additional flexibility, at a capital cost of well over $75 billion dollars.[2]

The components exist to modify supertankers to contain modern, 1,200-megawatt, gas-turbine power plants, along with AC/DC converters and CO_2 sequestration capabilities. We estimate that the cost would be about the same as FPSO vessels. One thousand miles of

HVDC cable to carry the electricity to multiple shore locations could be laid with a traditional marine cable-laying ship, for an estimated capital investment of about an additional $1 billion.

Fuel-efficient gas turbines with proven reliability in marine environments can be used to provide the high-density power plant. Modularization would be used to provide enough mobility that components requiring overhaul could be off-loaded for repair onshore. This modularity, combined with a modest amount of redundancy, has been demonstrated to deliver close to 100% availability in onshore gas-turbine power plants. Redundancy in AC/DC converters and HVDC cables could produce similar percentages of availability offshore, while providing real options for redeployment among newly emergent markets, given the rerouting and splicing potential of flexible HVDC cable laid on the seafloor. A hub-and-spoke design is envisaged for collecting the gas for powering the FGSO generators. CO_2 would be captured and returned at the gas fields for enhanced oil recovery (EOR) and sequestration tax credits (fig. 10–4).

NPV-based investment decisions create biased and uncertain, forecast-dependent valuations, especially for project many years out. Additional trade-offs, from the exercising of real options along the path to construction and deployment, can enable larger capabilities to profit when uncertainties solidify in the future. In other words, incremental investments in building the *capacity* to profit from the real options create a dynamic plan with decision gates for ongoing evaluation of capital investments. There is always more certainty of payback at a later date from any plan assembled today, and that advantage can be used to create a valuable, real-options framework for your company. Of course, real options for those investments must truly exist.

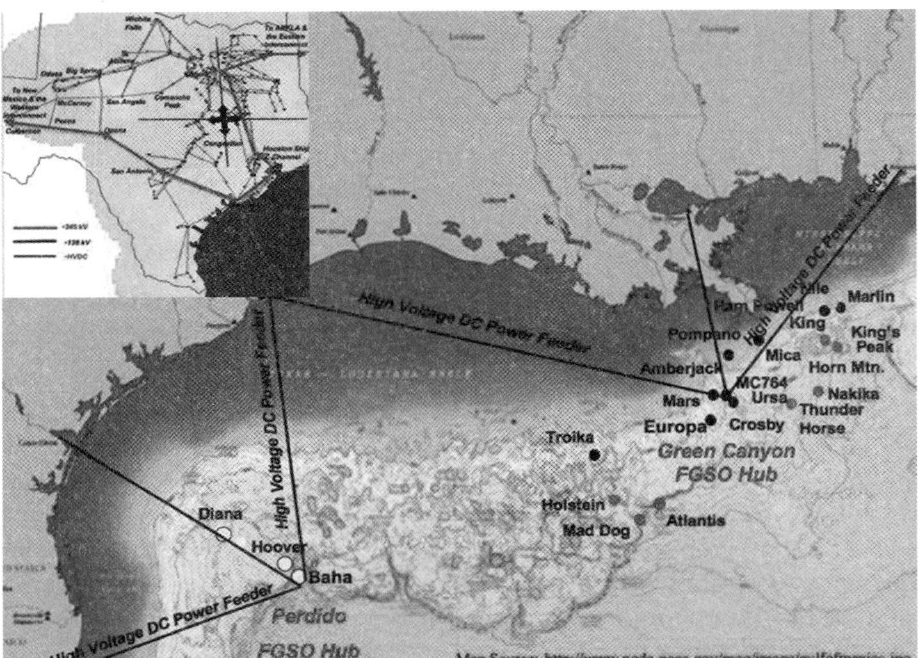

Fig. 10–4. Concept of western and eastern Gulf of Mexico ultra-deepwater gas and electricity hubs

We have used an American options model to estimate the value of maintaining real options for construction and deployment of ultra-deepwater FGSO hubs. We have then compared capital expenditures, real options, and strategic value with the traditional, NPV method of evaluating a five-step investment in the construction and deployment of an FGSO power production hub in the northern Gulf of Mexico (fig. 10–5). Such real options are valid when management has the ability to decide if each successive stage can be implemented after the results from the previous stage are finalized. In such options, only the previous sunk cost is at risk at any stage.

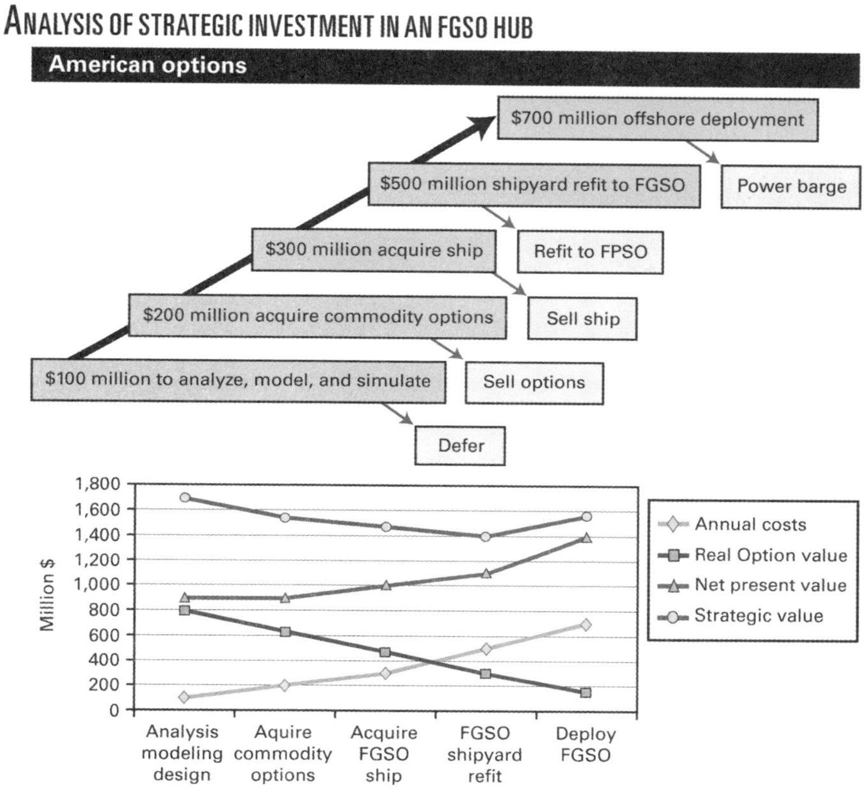

ANALYSIS OF STRATEGIC INVESTMENT IN AN FGSO HUB

American options

$700 million offshore deployment
Power barge
$500 million shipyard refit to FGSO
Refit to FPSO
$300 million acquire ship
Sell ship
$200 million acquire commodity options
Sell options
$100 million to analyze, model, and simulate
Defer

Million $
1,800
1,600
1,400
1,200
1,000
800
600
400
200
0
Analysis modeling design
Aquire commodity options
Acquire FGSO ship
FGSO shipyard refit
Deploy FGSO
Annual costs
Real Option value
Net present value
Strategic value

Fig. 10–5. Analysis of the real option value from investment in an offshore FGSO gas-to-electricity hub

The *strategic value* of being able to defer investments—to wait and see, once more information becomes available—is the sum of the real option value and the NPV of the asset, with cash discounted at the *hurdle rate* (the minimum rate of return that must be met for a company to undertake a particular project) and with volatility estimated.

If, over time, the volatility decreases, then uncertainty and risk are lowered, and the real option value decreases. The *cost of waiting* is evaluated by computing the dividend rate as a percentage of the asset value. The balance between collecting more information and the cost of waiting is continuously recalculated, so that the optimal strategic value can be maintained along the way. The ability to wait, while simultaneously preparing all that is necessary to execute, is worth several hundred million dollars in the ultra-deepwater FGSO hub in this case study.

Scenario analysis

Real options combine naturally with scenario analysis. In figure 10–6, we demonstrate this capability by evaluating the profitability from the execution of the FGSO real options described previously in a world in which gas and electricity prices are independently fluctuating at the same time that green CO_2 credits are being contemplated and congestion makes delivery even of abundantly supplied gas and electricity difficult. The value in these scenarios ultimately arises from the differences between natural gas and electricity prices (fig. 10–6).

We assume that price volatility is driven in the future by annual variations from summer to winter and occasional natural disasters, such as Hurricanes Katrina and Rita. Even black swans can be simulated as huge step-changes in the markets even if the reason goes unspecified, though we did not add that to our scenario analysis presented in figure 10–6. Our scenarios were then compared with recent variations in commodity prices over a three-year period in the mid-2000s. Prices at that time mostly fell within the *cheap-electricity-and-gas* scenario (lower left of fig. 10–6). The spark spread in the late 2000s migrated squarely into the *gas-expensive/power-cheap* quadrant (lower right of fig. 10–6). The upper-right quadrant of figure 10–6 represents the *green* scenario that the world may be headed for since global climate change will likely result in a carbon surcharge on current prices, *making both gas and electricity more expensive*. International shifts to massive renewable energy and away from the use of hydrocarbons might drive the spark spread into the *expensive-power/cheap-gas* quadrant (upper left of fig. 10–6). All four of these scenarios are feasible in the future. Real options combined with scenario analysis hedge a company through such uncertain times.

Price Scenarios for the Future*

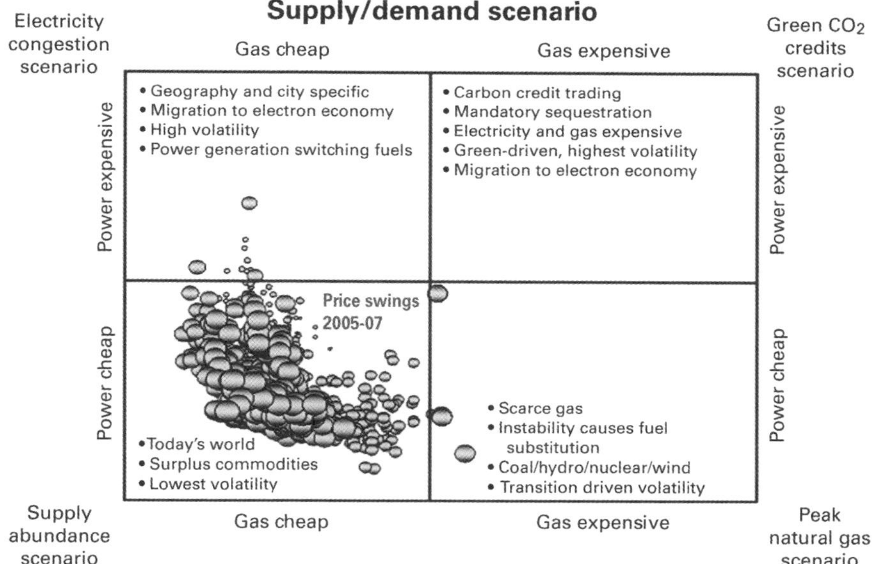

*Tracking of continuous price variations (real and futures) between electricity and natural gas for the Texas Gulf Coast from June 2005 (smallest bubbles) to December 2007 (largest bubbles).

Fig. 10–6. Scenario analysis of power and gas price variations from 2005 through 2007. Swings into any of the quadrants are possible in the next 20–30 years.

Harvesting of arbitrage between gas and electricity prices with either positive or negative spark spreads can be managed by the use of RL controllers during offshore operations. Significant profits could then be ensured from marketing a flexible mix, rather than selling a rigid flow of either gas or electricity, into markets represented by these highly variable price quadrants.

We foresee significant opportunities for carbon credits and an electric economy to develop as energy usage for transportation and heating both migrate to electricity. This is already occurring, with new technologies like plug-in hybrid vehicles and ultra-efficient electric heat pumps morphing with the growth of air conditioners worldwide.

CO_2 sequestration offers additional real options right at the source of ultra-deepwater production, to enhance both production through CO_2 flooding and profit from trading in carbon credits. There is a significant environmental upside from such real options. Given the assumption that the value of CO_2 sequestration would be high for the green scenario, returns from this investment are calculated to be far above the expected cost of capital.

Consider the value of carbon sequestration at the site of ultra-deepwater production should a carbon tax come into existence. Table 10–1 compares offshore enhanced oil recovery (EOR), using the CO_2 captured by the FGSO hubs, with estimated costs for onshore carbon sequestration.[3] Injection at the source of ultra-deepwater production could add a premium of $10–$20 per megawatt-hour equivalent in our scenario, assuming that the profit range is $3.50–$4.50 per ton of CO_2 injected (current costs in West Texas).

The real-options valuation needs to be combined with estimated revenue and operational and capital expense costs into a profit/loss model to determine the investment quality of the venture (fig. 10–7). We compute a payback period of five years for capital investment in the FGSO hub, assuming that an ultra-deepwater production facility is connected to both a gas pipeline and an HVAC/GSO electricity/sequestration hub. Without the carbon credit of $20 per megawatt-hour, the payback time would be seven years.

Table 10–1. Comparison of offshore EOR versus estimated costs of onshore carbon sequestration

Comparison of offshore CO2 injection for Enhanced Oil and Gas Recovery versus Carbon Sequestration onshore

Carbon Sequestration	Profit Range	Remarks
Injection of effluents from Gas generation offshore into deepwater reservoirs	***3.5 to 4.5 US$/tCO2 injected***	***CO2 has never been available for deepwater injection for Enhanced Oil Recovery. Little doubt CO2 injection preferable to water flooding.***
Carbon Sequestration	*Cost Range*	*Remarks*
Capture from a Coal-fired Power Plant	***50-75 US$/tCO2 captured***	***Net costs of captured CO2 compared to same plant without capture.***
Capture from hydrogen, ammonia or gas processing	***5-55 US$/tCO2 captured***	***Applies to high-purity sources only.***
CO2 pipelines	***1-8 US$/tCO2 transported***	***Per 250 km of pipeline at 5-40 MtCO2/yr.***
Aquifer storage on land	***1-8 US$/tCO2 injected***	***Excludes monitoring and verification.***
Ocean Storage	***5-30 US$/tCO2 injected***	***Includes transportation offshore 100-500 km.***
Mineral Carbonation	***50-100 US$/tCO2 mineralized***	***Includes additional energy needed for carbonation.***

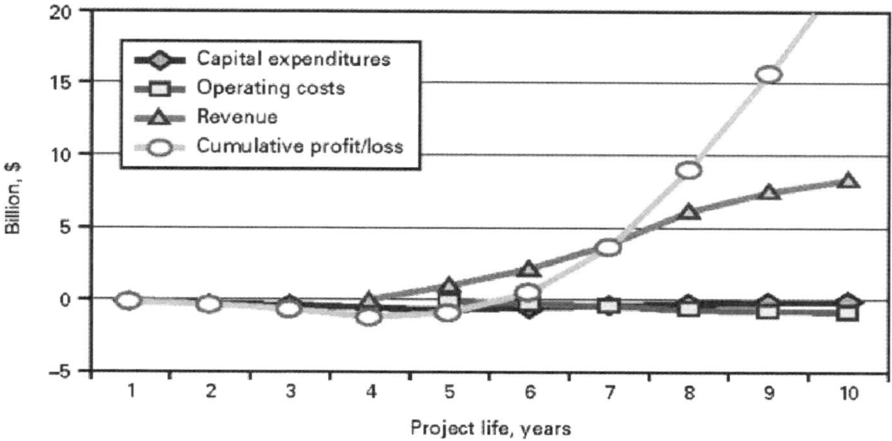

Fig. 10–7. Profit/loss model for real-options valuation. The investment into negative capital loss before first gas or electricity sales eventually drives the project to massive profitability.

Given long-term forecasts that are guaranteed to be wrong, a CALM methodology that pairs scenario analysis with real-options analysis can provide a powerful means of exploring alternative investment options under uncertainties in dynamic and market forces. The scenario analysis provides investment stages for real-options evaluation so that alternatives for deployment of capital can be identified into the future. The economic model for evaluation of the ROI for ultra-deepwater electricity production hubs using FGSO vessels and undersea HVDC power cables requires recovery of capital costs and sufficient future market flexibility onshore to sustain profitability for the life of the hub offshore.

MISALIGNMENT OF INCENTIVES

The outlook for making successful risk/reward investments in maintenance, system reliability, and other work-related portfolio management decisions is clear because the downside from big, bad events—black swans—can be catastrophic to a company's very survival. Put another way, proper maintenance of long-term assets—whether oil and gas fields, pipelines, LNG trains and terminals, refineries and petrochemical plants, transmission and distribution grids, or power plants—carries more safety risk than market risk, so the investments are more critical. Many times mistakes are made in the setting of performance metrics and incentives for top management of business units that leave the door open to trade-offs between HSE needs and other internal actions that will produce more near-term financial performance. The alignment of goals with performance metrics becomes intertwined with the integrated strategy and budgeting process of the company. Here, we show that planning, building, and operating a successful company requires not only lean processes and techniques but also drivers for personnel incentives that must be aligned with lean principles.

Energy companies can develop processes and techniques that are excellent operationally, yet they can fail to realize that these same tools can be used to propagate consistent HSE goals both inside and outside the company. An oil company that is very good at process management has recently had severe problems with bad safety and environmental events because operational goals and personnel incentives communicated to employees inside the company were at odds with communication of their HSE message to customers and shareholders.

Lean LNG project

We participated in a case study of the processes and controls for an exemplary lean transformation of the design, construction, and operation of a multibillion-dollar LNG project. We found that while the methods and techniques put in place by the project for design and development were outstanding, the incentives subsequently set up to propagate the lessons learned to global operations by corporate headquarters were out of alignment not only with the very lean management policies enacted by the LNG project but with corporate HSE goals. These HSE goals were themselves exemplary, designed to transform the environmental and safety profiles of the company that were presented to both Wall Street and the general public. We found, however, that the misalignment, propagated throughout the company—from the CEO to top-, middle-, and lower-level operational managers—brought disastrous consequences, including explosions and loss of life, severe environmental damage, potentially criminal activity, and a major negative impact on the global economy in which they occurred.

The upstream (exploration and production), midstream (transportation), and downstream (refining and marketing) segments of the oil and gas industry are all struggling to improve their processes. We studied the *capital value process* (CVP) as a showcase implementation on all those fronts via the huge LNG project, oil companies and then tracked the transfer of lessons learned throughout the rest of the company.

We found that in the execution of CALM principles and practices in the LNG project, all groups performed admirably, including commercial, subsurface, drilling, facilities, and operations. Interfaces and responsibilities between core and extended IPTs were well established, including subcontractor and governmental relations. One reason for this success was that the IPT members included governmental agency, partner, and contractor participation. Detailed facility plans were executed for the following: engineering, procurement, fabrication, commissioning (onshore and offshore), transportation, installation, hookup, platform modifications, and start-up. There was near-perfect alignment of incentives with key project drivers, including HSE impact, governmental relations, cost and value measures, schedule, and quality control. Risks were minimized and the organizational structure was clearly propagated to all concerned.

Three trains of LNG plant processing are operational, and potential natural gas shortfalls in Europe have been averted (so far) by LNG supplied from this facility. No accidents, no harm to people, and no damage to the environment have occurred at the new project site. It is the greenest project worldwide for the company. Drilling mud, cuttings, and produced water are reinjected into the geological reservoir, for example. Emergency upsets have all been successfully handled, rather than vented to the atmosphere. Power was centralized using turbine drivers that produced a 50% CO_2 reduction over that generated by comparable plants. Solar power was used extensively within the facility. The company used a lean CVP that deserves much of the credit for this admirable performance (fig. 10–8). CVP achieves operational excellence by providing a consistent, structured approach that improved capital efficiency through the continuous review and the challenging of assumptions.

Disconnect from lessons learned

Lessons learned from the project were immediately transferred to relevant organizations of the global oil company. However, a problem immediately arose with executive incentives. Top managers at business units set their own goals for the percentage of performance improvement of their divisions year over year. Since their personnel bonuses depend upon meeting these goals, the norm was to set the goals at a level that they were certain they could achieve. Unfortunately, cutting maintenance costs almost exclusively accounted for such sure things in the short run, principles that the lean CPV emphatically does not support.

The problem is that maintenance and supplies are the easiest costs to cut, but such cuts expose each asset to more in-service failures from excessive corrosion and wear and tear in the future. Thus, the intent of the CVP to spend money wisely to ensure profit for the life of the asset was subverted by the personnel incentive program focused on near-term financial performance metrics. The results were as predictable as they were tragic: pipelines went uninspected and ruptured, dumping millions of barrels of crude oil onto pristine land and sea; a refinery exploded because of deferred maintenance, killing dozens of workers; and company traders were arrested after it was discovered that they were meeting their performance goals through illegal use of insider-trading information.

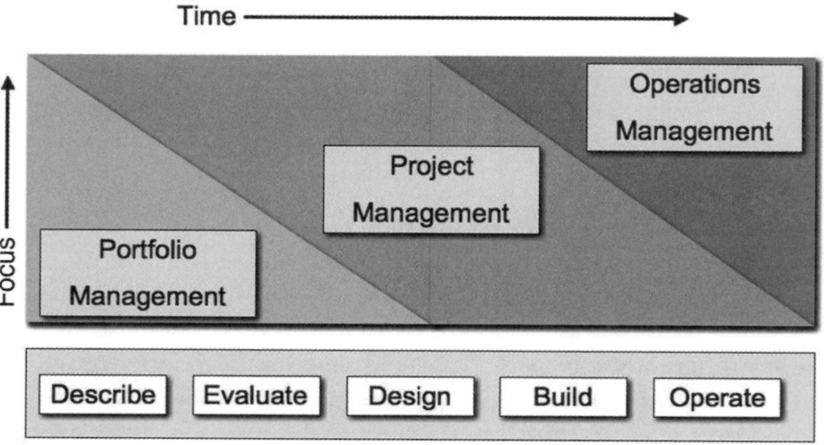

Fig. 10–8. The CVP, using peer reviews, peer assists, project HSE reviews, cost and schedule risk reviews, external benchmarking, risk and uncertainty assessments, technical integrity, and internal audits to deliver ahead of time and under costs, as the project progresses from left to right over time

The resulting damage to the company far exceeded its cost savings. In fact, the amount of money paid out in executive bonuses exceeded the costs that were cut. Aside from the lawsuits stemming from the pollution, loss of life, and criminal activity (estimated to be in the billions of dollars), the company's share price on Wall Street dropped to half that of its peers. The chief executive officer was fired, the chairman of the board of directors was forced out, and new management was brought in to clean up the mess. They immediately asked, "How can we prevent such an immense disconnect between our goals and our actions on our watch?"

The misstep between the performance goals of management and the incentive program designed to get the company's personnel to meet those goals should not happen in a CALM company. Why not? Because silos of isolation are prevented from forming by the continuous analysis, scoring, and feedback required within the brutally empirical framework CALM software enforces. In particular, preventive maintenance programs are valued within the portfolio at their real option value, rather than at their cost.

WELLNESS

CALM promotes development of tools and techniques for setting correct HSE policies in all its core operations and actions because they have such dramatic effects on all stakeholders in the company. In the *wellness* example that follows, we show how the health of employees can potentially be input into the ISM and simulations conducted to optimize employee well-being. Think of the wellness profile of each employee as inputs into an HSE SCADA system. Decisions can be made that have enormous impact on future performance on the basis of analyses of these data. For example, Con Edison required its employees to wear protective "white suits" and air-filtration devices while working at Ground Zero after the World Trade Center towers collapsed on September 11, 2001. Consequently, there have been few cases of subsequent lung disease among Con Edison employees. The understanding and proper mitigation of such operational risks are of paramount importance to the sustainability of any lean firm.

A pilot project was conducted to test whether the application of ML and statistical analysis techniques could reveal anything not already known about employee health as it related to perceived job stresses. The study had two goals: (1) to identify combinations of factors related to stress on the job that trigger sick days; and (2) to identify potentially unknown stressful-job categories that result in many sick days. The knowledge derived from this project might lead to stress-relief measures and monitoring systems that would improve employees' general wellness and job satisfaction.

ML analysis

We analyzed the existing databases that reflect the health, sickness patterns, and commuting activities of energy industry employees over the past 10 years. Our mission was to find previously unrecognized patterns of sickness that might be associated with job-related stress, to improve the general wellness of employees, and/or to detect situations requiring closer monitoring.

We used two kinds of data sources: *Static* data were used that represented background information about employees, such as age, date of hiring, job assignments, and departments. From these data, we also computed another community attribute, distance to work. Additionally, *dynamic* data were used that reflect the specific work activity of each employee on every day for over two years. We translated the dynamic data into a list of 12 additional attributes: time traveled, overtime worked, nights worked, training time, light-, medium-, and heavy-duty days, medical leave days, on-the-job injury days, hours spent at work in the field, personal days, and sick days.

Our initial feasibility study of the wellness program for HSE demonstrated that the goals of the pilot are indeed attainable. Boosting and SVM algorithms ranked employees by susceptibility to stress-induced sick days using data sets and attributes that captured information about each employee's background (age, department, distance to work, and type of field work), as well as past sickness history (fig. 10–9).

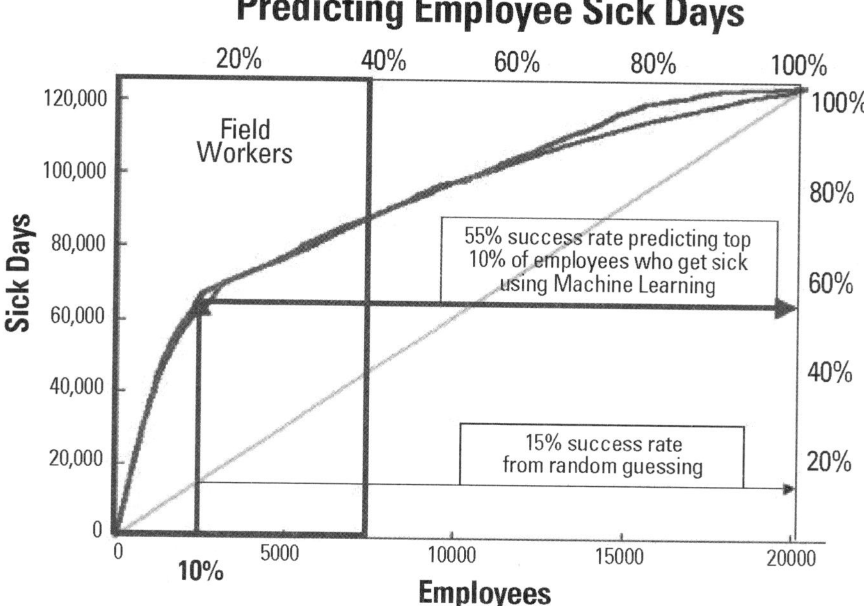

Fig. 10–9. ML methods used in ranking of susceptibility to sickness from job-related stresses. Employees were ranked from 1 (most likely to get sick) to 20,000 (least likely) by use of ML training, based on the number of sick days each was likely to require. We used both SVM and Boosting algorithms and then successfully blind tested the resulting ranking against a data set into the future.

We found several specific field assignments that were particularly prone to stress-related health problems. We ranked these job classifications from light to heavy workloads (from left to right, respectively, in fig. 10–10) and mapped the correlation with sick days for each of these jobs over 120 weeks (from top to bottom in fig. 10–10).

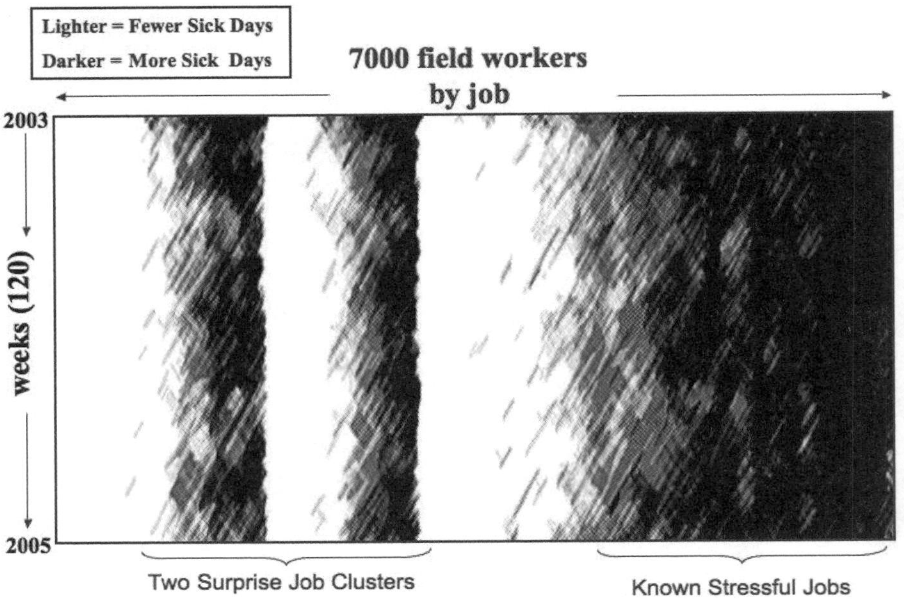

Fig. 10–10. Health risk for field jobs. Field work was classified from light work (left) to strenuous work (right) and was then mapped over time (progressing from top to bottom). Light vertical striping is used to associate class of work and number of sick days.

The stressful-job indicators at the right in figure 10–10 are well known and mostly deal with proximity to and repair of dangerous components, but the two groupings of light work but high stress at the left and center were surprises. The obvious correlations between light- and medium-effort job assignments and sick days had previously been unrecognized. The company can now begin studies to improve stress conditions for employees undergoing what turned out to be surprisingly more strenuous working environments than understood by employees and management alike.

Small-scale variability in the results across time periods is also evident in the variations in intensity of the vertical shading in figure 10–10. In the future, fine-scale implementation can be set up, using ML techniques to validate whether policy changes that are variable from week to week can improve wellness within these particular high-stress job clusters.

The preceding examples raise the hope that ML analysis can be used to design better work environments for energy industry employees. The dedicated men and women that work in hazardous and stressful jobs throughout the energy industry deserve to leave safely and unharmed at the end of each shift.

CUSTOMER SATISFACTION

We close this chapter with *customer satisfaction* examples that show the critical connection between employees and systems that communicate with customers' real-world needs and desires. Higher customer satisfaction enables the potential for growth in many indirect ways. For examples, satisfied customers buy more products and services; reduce chances of degrading the company's intangible asset of reputation; improve employee happiness resulting in productivity improvements; promote regulator satisfaction; and reduce complaint calls and correspondence. Maintenance of customer satisfaction can be tenuous even in lean companies if attention is not continually focused on delivering the correct information to customers when product problems or service disruptions occur. Particularly in the energy business, your most important employees are those that deal directly with your customers.

The key principles for CALM consist of setting up a lean management structure and associated policies and incentives that build a foundation for continuous improvement through the healthy reliance on computational efficiency and communication of all needed information openly. Building this foundation starts with requirements for high-quality information about how your company works and ends with the development of an ISM of the entire business (which has yet to be fully implemented in the energy industry). The purpose of CALM is to increase enterprise value, and that requires a focus on producing happy customers who are satisfied with the products and services of the company. Therefore, CALM seeks to improve relationships with customers, and in many cases, that requires lean transformation of call-center actions and policies, since most of the interaction occurs with them when customers have problems. We discuss four examples in the following sections.

Call-center rules engine

Most call-centers utilize a rule-based customer information system (CIS) to assist and direct their customer service representatives (CSR). Built into the CIS application is some kind of rules engine (RE) to detect problems and inconsistencies with products or services (e.g., bills might be wrong or the lights or gas out). The RE then triggers issues for investigation and resolution by the CSR. In most cases, legitimate reasons exist for the billing irregularities detected or services interrupted, but these must be properly conveyed to the customers affected.

Most REs are not mathematically sophisticated enough to recognize patterns and learn optimal solutions. Goals of a CALM RE should be to make both CSR and customers happy through:

- Classification of recurring problem types in order to eliminate Sherlock Holmes events. These are repeated patterns of cause-effect-solution that computer intelligence can easily identify and classify, thus eliminating the need for the CSR to conduct deductive investigations and give-and-take with customers to try to discover the root cause of each repeated irregularity.
- Tracking of root causes as they occur and monitoring of the degree of success of each fix.

- Automated responses to as many solutions as possible are built into a best-practices database to learn and store effective solutions.

We have designing a lean RE system for call-centers in which chained matrices are used to indentify causes, predict effects, and deliver best possible solutions to the CSR for billing irregularities related to meter changes. Causes are related logically, temporally, and by physical location and proximity to crews that can fix the problem. Events predicted with varying certainty based on the pattern of causes are then prioritized by the RE on the basis of their benefit to cost ratio to the customer, not just the company. Solutions are sent from the RE to the CSR as action plans intended to alleviate the problems, prioritized according to the most likely successful outcome. The RE then tracks actions taken to validate and learn proper prediction skills. When a root cause is detected, the RE sends operations or planning a fix request. For each cause and effect, the best-practices database is populated with

- The problem description with the likelihood of its occurring again.
- List of likely causes that contributed to the problem.
- List of possible solutions to the problem and/or any of the contributing anomalies.
- Statistics of past successes or failures with each solution.
- Cost-benefit analysis performed for each suggested solution, again from the point of view of the customer.

Lost enterprise value

Software tools can help manage customer satisfaction by continuously measuring the economic threats posed by risks associated with perceived and real system failures. As in the illustration that follows, an energy company can compute its lost enterprise value based on the number of irate customers affected and the duration of their pain during a service disruption using these tools. Then, the business risks can be quantified and mitigated with better asset design and deployment plans. Put simply, system designs and restoration efforts can be better engineered and executed to reduce these risks to the enterprise if the true costs to the company are recognized.

Lost enterprise value can be simulated, and mitigation plans can be prepared before big, bad events—black swans—happen. For example, informing customers when service will be restored after a blackout has enormous value to the enterprise in terms of customer goodwill. An even better outcome would be to have simulated many, many scenarios of possible power outages, so that each new event can be quickly and accurately identified and the likely issues associated with the outage can be conveyed to customers as soon as possible.

Figure 10–11 depicts the realistic enterprise value lost from various scenarios of electrical outage and restoration efforts. The checkerboard surface represents the optimal value to customers at all points in time at which the outage is just over and electricity to most customers has been restored. Each thin curved line represents a scenario of restoration efforts for different kinds of outages; the shapes of the curves over time are a function of the ways in which responses are handled. The surface provides the means to analyze the effectiveness of changes to drivers of reliability, such as system design, including the value of efficient and accurate communication of the problems and likely durations of outages to customers.

In order to preserve reputational value, the utility needs to convey to affected customers that it clearly understands the expected impact that its performance will have on their quality of life. Having previously modeled the business processes and likely schedule for repair, the company can inform customers how the event will likely play out. An overnight outage communicated to 5,000 customers might prompt those customers to sleep at friends' houses or go out and enjoy other parts of the city. Lack of information to customers might result in a fearful night spent at home with the lights off and the doors locked.

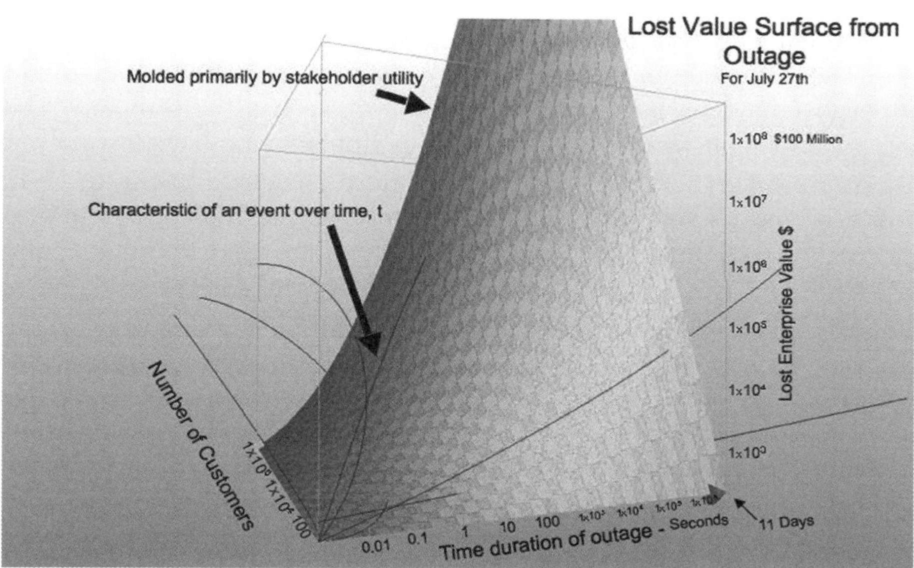

Fig. 10–11. Representation of how an electricity outage might affect erosion of enterprise value. The diagram can be considered four-dimensional, because the event surface changes continuously with time and the remediation effort.

Blackouts are bad

Another example of the value of customer satisfaction is supplied by a recent electrical network failure that blacked out 120,000 customers in a major American city. Even though the lights were out for only two hours, the media made the outage into its top story on all news channels over the next several days, interviewing customer after customer complaining that nobody was telling them what was going on. As the airlines know full well from their pitiful crisis performance in the early 2000s, a few complaints over the airwaves can metastasize into millions of people who are irate about the company's performance.

The CALM approach focuses on quantifying factors that might otherwise have been ignored by decision-makers in the company. The engineer evaluates only the number of customers who have lost power. By contrast, the CALM parametric surface represented in figure 10–11 could be used to estimate the restoration threat to the enterprise and calculate the number of affected customers beyond those actually blacked out. This customer-focused approach to enterprise value is important to meeting perceived expectations of service quality.

The duration of an outage increases the slope of the surface exponentially in figure 10–11. If the outage were limited to a small number of customers or were resolved quickly, then lost enterprise value would be significantly lessened for each incident. That way, the true value of rapid restoration—and of limiting the outage spread—can be weighed against the added cost of overtime and logistical staging. The CALM methodology aims to measure, from a customer's perspective, the design needs required to satisfy the happy customer beyond the pure electrical engineering methodologies of the past.

Each circumstance differs, of course, but modeling and measuring possible outcomes through the simulation of scenarios before events happen is a hallmark of CALM. Quantification requires utilizing knowledge of the existing assets and services used (i.e., the as-is case), in order to develop alternatives that optimize value to all stakeholders in the future (i.e., the to-be case).

Overbuilding

Consider another example of a significant opportunity to improve customer satisfaction through better asset utilization that should be able to lower rates in the electricity economy of the future. The load behavior of New York City from 2004 to 2006 is shown in figure 10–12.

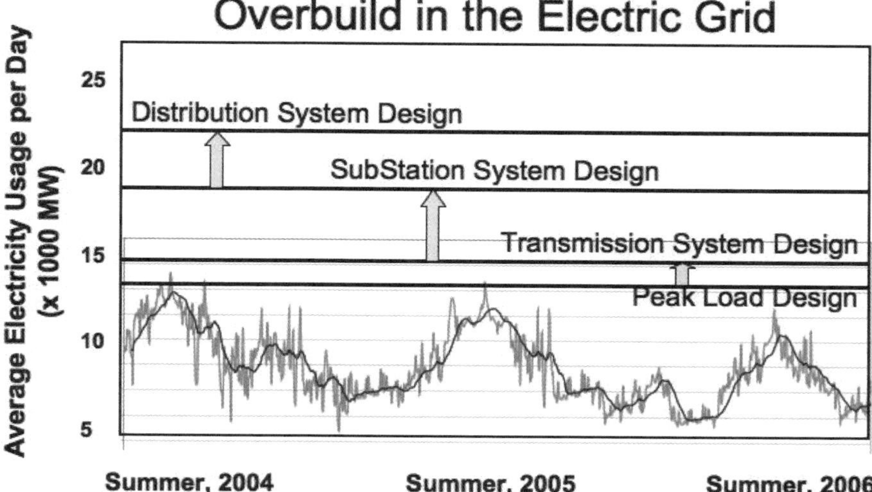

Fig. 10–12. Overbuild in the electric grid. Overbuilt infrastructure is a characteristic of most urban utilities. Actual load over three years is shown, over which the peak load design was touched only 2.5% of the time. Only a quarter of the distribution system design capacity, half of the substation system design capacity, and less than three-quarters of the transmission system design capacity were used (top). This underutilized, overbuilt capacity is present both in the city grids and in the suburban territory of most urban areas around the world.

The overbuilding of capacity evident in the fact that designs are all far higher than peak usage holds true for most utilities in the world. This design overbuild takes into consideration the potential of a massive, once-every-hundred-years heat wave. The average load for any hour is way below the design peak load. However, that's not the only measure of underutilization. Utilities also add layers of contingency—in the form of redundant transmission lines, generators, substations, primary feeders, network protectors, transformers, and secondary cables—to protect against the failure of these vital components on the highest-peak-load day. For example, a fully loaded, five-transformer distribution substation in Manhattan is, at best, using 60% of its capacity even on the peak-load day. There are many distribution substations countrywide that are utilized at only about 30%

of their capacity, on average, for the year. While CALM methodology supports some overcapacity in design to allow downtime for repairs, its enhanced use and reliance on computer intelligence would enable much more efficient operations that are closer to design specifications. Customers will ultimately be happier because rate case increases for major capital investments to sustain the overbuilding can be avoided, and power delivery will be just as reliable because of the increase in intelligence of the grid.

Most of the data required for such CALM transformations in interactions with customers already exist but are rarely used for decision-making purposes, except perhaps for incident investigation or management-initiated special studies (fig. 10–13).

All six horizontal boxes in figure 10–13 have the potential of improving enterprise value and customer satisfaction simultaneously. The first box, "Higher Human Diversification of Services," means that user-friendly software tools assist workers in performing a wider variety of job assignments than ever before. The goal is to drive toward zero specialists. The second box indicates higher quality as a means to improve shareholder value, assuming that a satisfied customer always adds enterprise value. The third box strives to lower costs through zero waste. Again, there is a fine line between waste for one stakeholder and value to another that must be taken into consideration. The fourth box, "Customer Service Satisfaction," is another overarching goal obtainable through timely utilization of customer information software. The fifth box is a goal that is normally forgotten, which is love of the job. People have a natural desire to be proud of their work and comfortable with their surroundings. They want to have positive relationships with the people they communicate with on a daily basis, whether fellow employees or customers, and humans really do strive to help others.

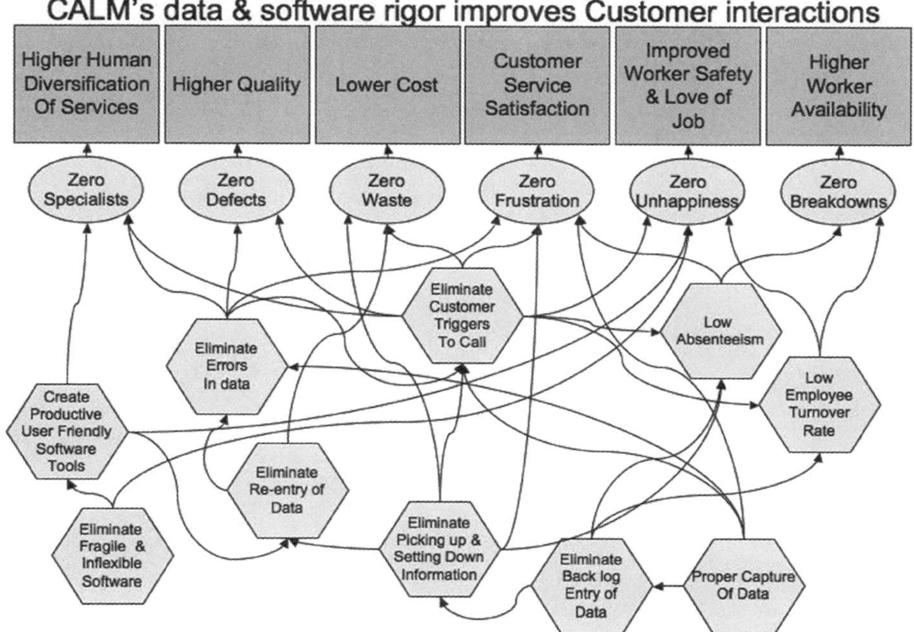

Fig. 10–13. A cause-and-effect diagram to indicate goals that will enable a company to improve enterprise value through better customer interactions via a transformation to CALM principles

When the management, processes, and systems at work make this work environment possible, people will be more productive and happier. Feeling good about oneself promotes a healthy lifestyle that can reduce accidents and lower health costs. These basic principles create a win-win-win between customers, company, and workers. All six of these goals can be achieved through improvements in processes and systems. For employees who deal directly with customers, migration from purely cost-based performance metrics to quality-of-response– and quality-of-resolution–based performance metrics will help to modify considerations of customer satisfaction from the cost-center mentality that it is predominant today.

NOTES

[1] International Panel for Climate Change, United Nations. 2007.

[2] *Maritime Reporter,* August 2003.

[3] These examples represent the work of the authors and scientists from the Center for Computational Learning Systems (CCLS) at Columbia University. David Waltz is the director of CCLS. Marta Arias did the research in the "Wellness" case study.

11 ENERGY FUTURE

We live in a biosphere that is stressed. Our current, excessive use of Earth resources is unsustainable. We extract and burn energy stored in fossil fuels and send back harmful pollutants and climate-changing CO_2 into the environment. Notably, CO_2 is neither a pollutant nor a poisonous gas, but provides breath for all the plants on Earth. Unfortunately, accumulation of this molecule in our atmosphere causes the *greenhouse effect,* trapping the sun's heat in our atmosphere and causing global climate change. Temperature and CO_2 concentrations have dramatically increased since 1900, but the implications for the future are uncertain (fig. 11–1). What is certain is that we are conducting a gigantic experiment with the climate of Earth, and no one knows how it will turn out. Ironically, the melting of the polar ice sheets might even trigger global cooling and another ice age (since we are in an interglacial cycle).

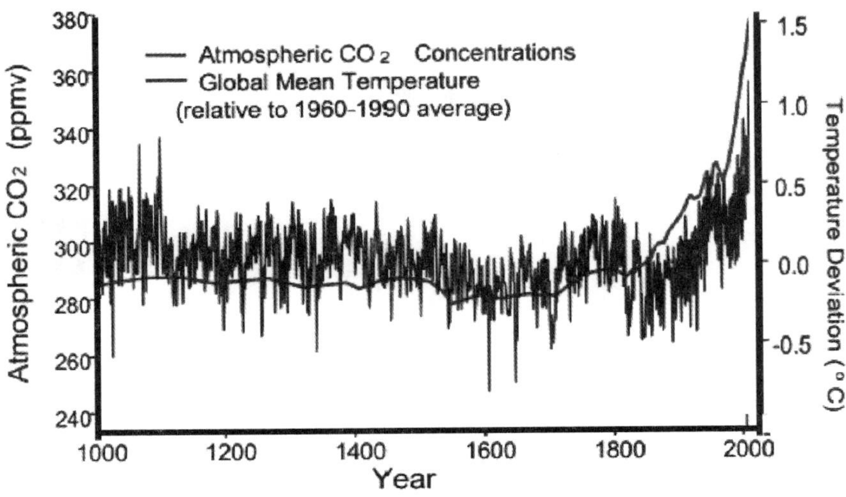

Fig. 11–1. Increasing CO_2 in our atmosphere prior to the rise of global temperature (***Source:*** Basic Research Need for a Hydrogen Economy, Report of Department of Energy Basic Energy Sciences Workshop on Hydrogen Production, Storage and Use, 2003)

The primary cause of this degradation of our environment is the noble pursuit of the energy to provide us with clean water, sanitary and safe living conditions, food for our tables, global mobility, and comfort. Because of our human drive toward the pursuit of happiness, the energy industry has been paid to take massive volumes of natural resources from Earth to deliver power to society. In this chapter, we discuss impediments to energy sustainability and potential solutions that CALM can provide through enablement of a global electric economy. CALM provides a methodology to achieve the smart electric grid to support such a transition.

Earth will be fine 100 million years from now, just as it was 100 million years ago, even if humans disappear tomorrow in a global environmental disaster. Nothing we can do is as bad as the natural catastrophes that the planet has experienced in the past. Eighty million years ago, a meteor the size of Dallas, Texas, wiped out the dinosaurs, and half of all life on the planet. We cannot permanently mess up the biosphere to that degree. Humans are not powerful enough. Sure, we may change the climate, but not to the same extreme as occurred more than 300 million years ago, when all the ice on the planet melted because several natural cycles in the revolution and rotation of Earth aligned themselves. The ocean covered the entire midlands of North America (think beachfront in Calgary, all of Central America south of Mexico City under water, and only the Appalachian and Rocky Mountains above water). During the last glacial period—a mere 18,000 years ago—there was no Long Island or Cape Cod off the East Coast of the United States; instead, there was a mile or more of ice on top of all of North America, from the North Pole to Nebraska and eastward to New York City. Nevertheless, an electric economy is emerging that can significantly improve global energy sustainability.

In this chapter, we present our vision of the electric economy of the future, which uses the tools developed through CALM to provide the intelligent solutions that human society needs to power the next century and beyond.

The following are some of the current constraints to better intervention to slow the effects of energy use on global climate change:

- *Lack of local measurement of sources.* The fundamental, lean principle of *define and measure* is clearly an issue. A good model of the impact of global climate change on the environment has been developed by the National Aeronautics and Space Administration Goddard Institute for Space Studies. Their model computes the footprint of carbon and other greenhouse gases on a global scale. However, it is difficult to relate the observed environmental change from a *causal* perspective without the local footprint of energy sources and sinks. What's missing in this model are the direct local contributions that are required in order to prevent the expansion of this carbon footprint over the next century. We simply do not know the specific sources factory by factory and town by town. We don't know how much CO_2 is being generated from producing and delivering any specific product, whether it be toys, cars, or air conditioners, and from plants in China or India, versus the same product manufactured in Europe or the United States. We have only generalities rather than maps of the volume, time-variance, movement, and so forth, of the CO_2 plumes produced by cities like New York at the moment.

- *Decision-making under deep uncertainty.* Real options and similar methods, like complexity analysis work in decision-making under conditions of deep uncertainty, can address how to derive robust policies in a methodological way based on simulations that incorporate large uncertainties (e.g., Lempert, et al, *Shaping the Next One hundred Years*, Rand, 2003).
- *Disclosure and auditing of the environmental footprint.* There will always be mistrust that others are taking advantage of the environment to serve their own self-interests. After all, the exploitation of resources that has been occurring for the past two centuries is largely responsible for the recent advancements of humankind and two world wars. As we consider global social responsibility, can corporations and governments be trusted to make auditable disclosures of the environmental footprints of their economies? Whoever steps up first will likely reap the benefits of designing the remediation of that footprint through tradable allowances and technological advancements. The European Union appears to be leading the world in climate change at the present time. A neutral auditing organization that is globally accepted and has the authority to provide effective refereeing and standardization should be created to enable full disclosure. Perhaps the International Energy Agency is the model for a future International Environmental Agency. Better yet, the CALM solution would be to create one global ISM housed at the new International Agency for Energy and the Environment—they are inseparable, after all.
- *Price for controlling environmental degradation.* The products and services sold today are not priced on the basis of recovery of the additional cost to repair environmental damage from their production and distribution. Being environmentally neutral is costly. Perhaps it makes sense to have a market-based pricing

> mechanism with trading similar to sulfur dioxide (SO_2) and nitrous oxide (N_2O) allowances that are currently in place in power generation. Charging a consumer tax or some other end-use fee mechanism on emissions from primary fuels would encourage innovative ways to efficiently drive down carbon emissions. Then, individuals could make informed decisions about the wise consumption of products and energy use, including choosing distributed renewable sources (e.g., green power from solar and wind generation) and all-electric cars over hybrids.

We foresee that the development of an electric economy will not only speed the removal of these impediments, but become the principal enabling technology for allowing both energy and environmental sustainability for the Earth. As Rick Smalley, the Noble Prize–winning nanotechnology scientist, repeatedly pointed out in the early 2000s, if Earth had cheap and abundant energy, most other bad things that happen to people on the planet could be mitigated. We could

- Purify all the drinking water
- Produce all needed food
- Clean up the environment
- Prevent global climate change
- Eliminate poverty and disease
- Educate every man, woman, and child on the planet

Sound like a huge overstatement? The following is his Nobel vision:[1]

> Energy is the single most important challenge facing humanity today. As we peak in oil production and worry about how long natural gas will last, life must go on. Somehow we must find the basis for energy prosperity, for ourselves and for the rest

of humanity, for the 21st century. By the middle of this century, we should assume we shall need to at least double world energy production from its current level, with most of this coming from clean, sustainable, CO_2-free sources. For worldwide peace and prosperity, it must be cheap. We simply cannot do this with current technology. We will need revolutionary breakthroughs to even get close. *Electricity will be the key.*

Consider, for example, a vast interconnected electrical energy grid for the North American continent from above the Artic Circle to below the Panama Canal. By 2050 this grid will interconnect several hundred million local sites. There are two key aspects of this future grid that will make a huge difference: (1) massive long distance electrical power transmission; and (2) local storage of electrical power with real-time pricing.

Storage of electrical power is critical for the stability and robustness of the electrical power grid, and it is absolutely essential if we are ever to use solar and wind as our dominant primary power sources. The best place to provide this storage is locally, near the point of use. Imagine by 2050 that every house, every business, every building has its own local electrical energy storage device, an uninterruptible power supply capable of handling the entire needs of the owner for 24 hours. Since the devices are small and relatively inexpensive, the owners can replace them with new models every five years or so as worldwide technological innovation and free enterprise continuously and rapidly develop improvements in this most critical of all aspects of the electrical energy grid.

Today, using lead-acid storage batteries, such a unit for a typical house to store 100 kilowatt-hours of electrical energy would take up a small room and cost over $10,000. Through revolutionary advances in nanotechnology, it may be possible to shrink an equivalent unit to the size of a washing machine and drop the cost to less than $1,000. With intense research and entrepreneurial effort, many schemes are likely to be developed over the years to supply this local energy storage market that may expand to several billion units worldwide.

With these advances, the electrical grid can become exceedingly robust, since local storage protects customers from power fluctuations and outages. With real-time pricing, the local customers have incentives to take power from the grid when it is cheapest. This in turn permits the primary electrical energy providers to deliver their power to the grid when it is most efficient for them to do so and vastly reduces the requirements for reserve capacity. Most importantly, it permits a large portion—or even all—of the primary electrical power on the grid to come from solar and wind.

The other critical innovation also needed is massive electrical power transmission over continental distances, permitting, for example, hundreds of gigawatts of electrical power to be transported from solar farms in New Mexico to markets in New England. Then all primary power producers can compete with little concern for the actual distance to market. Clean coal plants in Wyoming, stranded gas in Alaska, wind farms in North Dakota, hydroelectric power from northern British Columbia, biomass energy from Mississippi, nuclear power from Washington, and solar power from the vast western deserts—remote plants from all over the continent can contribute power to consumers thousands of miles away on the grid. Everybody plays. Nanotechnology in the form of single-walled carbon nanotubes (a.k.a. "buckytubes") forming what Rick Smalley called the armchair quantum wire may play a big role in this new electrical transmission system. Such innovations in power transmission, power storage, and the massive primary power generation technologies themselves can only come from miraculous discoveries in science together with open competition for huge worldwide markets.

THE SCALE OF THE GLOBAL ENERGY PROBLEM

By 2050, the world will need 30 terawatt-hours per year to supply total energy demand (fig. 11–2).[2] In 2008, we used a little more than 16 terawatts, with 85% coming from burning hydrocarbons, and one-fifth of that was consumed in the United States. The greatest single challenge over the next few decades will be to supply this amount of energy for the 10 billion people who will live on the Earth then. At a minimum, we will need an additional 15 terawatts (double our present global consumption of energy). This is the equivalent of 150 million barrels of oil demand per day. In 2008, we produce 85 million barrels of oil per day. For worldwide peace and prosperity, we need this energy to be cheap and to be available from many more clean energy sources than we have today. Figure 11–2 points out a significant problem for the people on Earth. We have never experienced the scarcity in energy supplies that the dip in supply implies (arrow in fig. 11–2). We simply cannot deliver this amount of energy with current technology.

The 30 terawatts cannot be generated from oil, gas, and coal by 2050. The much-anticipated peak or (more likely) plateau in oil production in the world (fig. 11–2) will make it impossible for oil to supply enough energy for global demand by around 2030. We have never experienced such a dip in supply, except perhaps in the Dark Ages, when humans ran out of wood.

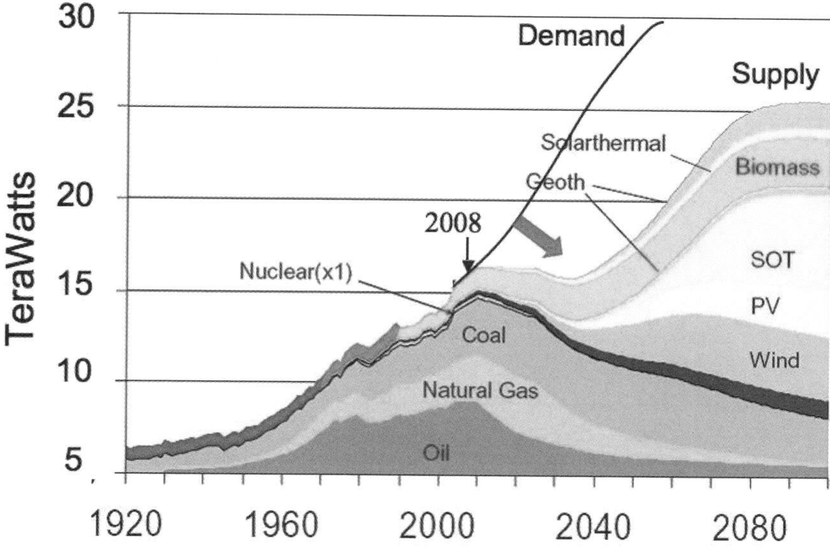

Fig. 11–2. Future supply of energy. In this view, oil and gas supply peaks within the next few years. PV: solar photovoltaics. (***Source:*** M. Zerta, P. Schmidt, C. Stiller, and H. Landinger, Alternative World Energy Outlook (AWEO) and the role of hydrogen in a changing energy landscape, 2nd World Congress of Young Scientists on Hydrogen Energy Systems, Torino, Italy, June 2007 [http://www.lbst.de/publications/presentations2007/LBST_HYSYDays_6-8JUN2007_Torino.pdf])

The problem becomes how to transition from the hydrocarbon economy of the 20th century to the electric economy of the 21st century. Only electricity can get us out of this box. Our use of electricity is terribly inefficient at the moment. *The Economist* estimated that smartening the electric grid to intelligence levels comparable to that of the Internet would allow the presently produced electricity to supply 50% more demand, all by itself.[3]

Unquestionably, we should come up with an environmentally sound source for this conversion to an electric economy, and that means reducing carbon emissions from power plants and using much more alternative energy that is produced from wind, solar, geothermal, and hydroelectric sources (fig. 11–3). Nuclear power is also a potential source of increased supply. The safety of nuclear fission technology has advanced considerably in the past decade or so, making it a viable choice for the future in many areas of the world; however, fission has yet to come up with a technological solution to high-level radioactive-waste disposal, and it represents a target for terrorism. Nuclear fusion as a cheap, abundant, and inherently sustainable electricity source has been 50 years in the future for the past 50 years. Geothermal, wind, and solar power have sufficient size but have significant costs, as well as technological and distribution problems.

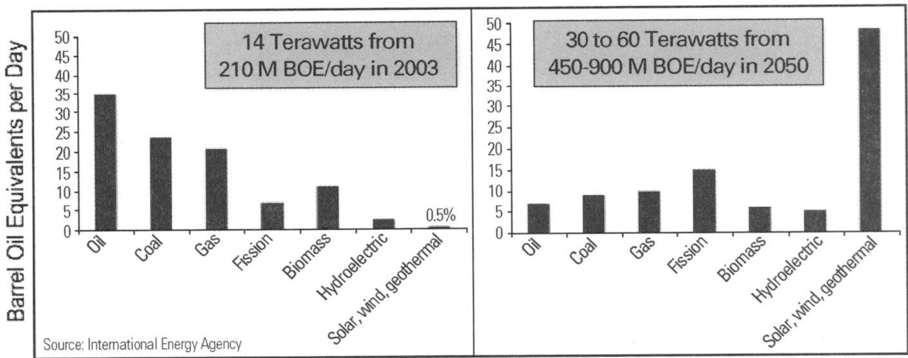

Fig. 11–3. Rick Smalley's terawatt challenge: to produce enough energy that it is cheap and abundant

The following list outlines Rick Smalley's anticipated world energy landscape in 2050:

- Electricity dominates the energy economy.
- Local distributed storage enables use of massive renewable resources.
- Abundant local distributed generation is available, primarily from renewable resources.
- Transportation is dominated by electric vehicles.
- The grid is smart, resilient, and can satisfy customer demand even while being less reliable.
- Primary power backbone of the electric economy will be via high-voltage direct-current (HVDC) transmission lines.
- Vast solar and wind farms, remote nuclear, clean coal, stranded-gas, wave, tidal, and hydroelectric power plants, as well as spaced-based solar power microwaved to Earth, will each operate on a terawatt scale.

All of the preceding sources are needed to meet 30 terawatts of demand each day.

The sooner we begin the transition to the electric economy, the closer we will come to realizing Rick Smalley's grand vision of cheap and abundant energy for all who need it. The following are some of the more pronounced impediments to the electric economy:

- R&D in most electric utilities is nonexistent, and the substantial investment in the oil and gas industry is primarily focused on production and marketing of traditional hydrocarbon products without much consideration of how to mitigate their environmental impact.
- Present regulatory oversight drives risk aversion in electric utilities, inhibiting the application of state-of-the-art technology. In fact, recent deregulation efforts in many areas of North America has increased investment uncertainty.

- Governmental funding for the advancement of the electric grid is virtually nonexistent worldwide (certainly in North America, where, e.g., Con Edison invests more in electricity distribution R&D than does the entire U.S. Department of Energy).
- An understanding of how the electric economy could be used to reduce greenhouse gas emissions is also almost nonexistent among environmental scientists and engineers.
- While most utilities lack the business capabilities to effectively create an intelligent electric grid design, the more process-savvy oil and gas companies have the wherewithal but have shown little interest.
- In sum, getting to the electric economy is a CALM problem, and energy companies don't understand the connection yet.

ALTERNATIVE ENERGY

Most renewable energy sources are intermittent, highly variable, and unpredictable (fig. 11–4). Giant solar power and wind farms may be located thousands of miles from metropolitan centers, but such sources cannot simply be added to the existing electric grid—it is not smart enough. The management of the grid will require digital control, automated root-cause analysis, and rerouting capabilities similar to those that keep the Internet running smoothly. In short, the present global electric grid will become unstable if the higher loads and unpredictable timing from massive, remote, renewable power sources are added.

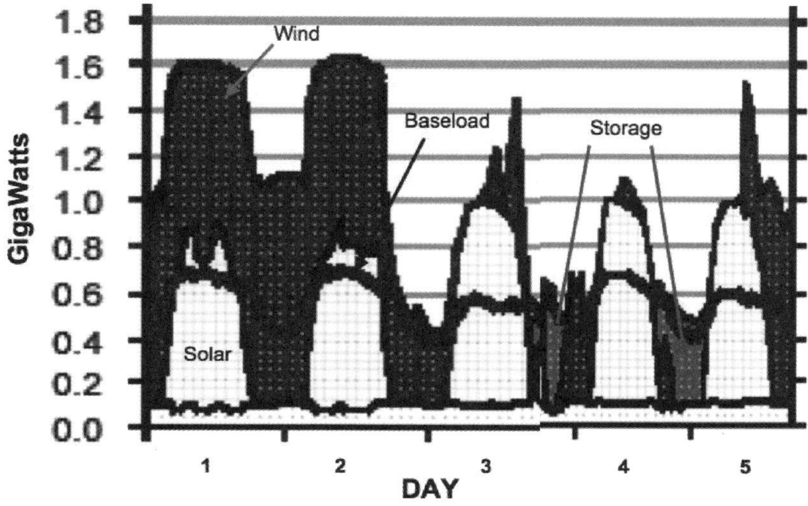

Fig. 11–4. Day-to-night variation in solar and wind power over a five-day period. Coordination with storage has the potential to stabilize the grid. (***Source:*** G. D. Berry and A. D. Lamont, Carbonless transportation and energy storage in future energy systems, in ***Innovative Energy Strategies for CO_2 Stabilization,*** edited by R. G. Watts, New York: Cambridge University Press, 2002)

In addition, the electric grid is not well equipped to deal with the large increases in congestion that are being stimulated by this long-distance electricity transport. The restructuring of electricity markets has promoted not the integration of the delivery system, but instead the partitioning of responsibilities. Problems such as a dramatic increase in blackouts and brownouts worldwide since 1998 will only get worse. The addition of unpredictable loads to the transmission grid will propagate disturbances more and more nonlinearly if we migrate to new continent-wide, supergrid environments because required decision speeds increasingly become too fast for humans to manage. As demand, sources of energy, complexity, and distances between demand and supply increase, this "greatest machine ever built"[4] will become increasingly vulnerable not only to blackouts but also to terrorist attacks and equipment breakdowns. Moreover, the weather is expected to become more violent and unpredictable because of global climate change.

It is clear that we will need to modernize the grid, but we must also make it smarter. The only alternative today is to double—or even triple—the hardware that makes up the grid. Particularly in our great urban centers, we are simply running out of room. Alternatively, we can better manage the current system that is overbuilt, by installing digital controllers, electronic switches, routers as in the Internet, and higher-capacity transmission lines. These will all be required if we are to install massive amounts of renewable energy sources within the next 10 years.

TODAY'S ELECTRIC ECONOMY

If the electric grid is to form the foundation for a transition from our present hydrocarbon-dominated energy world to a global electric economy, then the grid and the energy companies involved in building and maintaining it will need to both become smarter. Computer-aided intelligence is what CALM is about.

Solutions to address these energy supply and demand issues will require computerized management of diverse assets and resources within the electric grid. However, only through the development of business capabilities within the energy companies will applications of new technologies and software algorithms enable the efficiency increases needed for management of the electric economy. The ISM that we've spoken about throughout this book will provide this increased intelligence for the smart planning, engineering, design, building, maintenance, and operation of the electric grid of the future.

Mathematically advanced computational algorithms are needed to drive waste from such massive implementations of the electric economy. An ISM of the entire electric grid of a continent would enable each company that works within that umbrella to use the tool to learn how to best distribute electricity to its customers. Such a continental-scale ISM could optimize the reliability and efficiency of the grid, enable business optimization, and manage the complexity of new resources to maintain a stable operating environment of electric supply, demand, and delivery. The development of such an ISM would allow independent system operators or regional control-centers to manage much more complex configurations of the grid in order to maintain resiliency over long distances. However, it would probably be the largest machine ever built by humans—and certainly the greatest computer sciences challenge ever undertaken.

Through this solution, one might even imagine a distributed-storage/generation electric grid that spans the continents (a global vision of Buckminster Fuller). For example, suppose that all energy were to be transported as electricity over next-generation nanotechnology wires, rather than by the current physical transportation of mass (coal, oil, or gas by pipeline, ship, and train). Suppose also that the electric grid had vast, continental-scale interconnections, with millions of asynchronous, local storage and distributed-generation sites built into it. The entire system could be continually innovated with new smart controllers, and hundreds of new power plants would *not* have to be built in the first place (fig. 11–5).

Fig. 11–5. Smart controllers. Millions of grid-friendly controllers distributed throughout the electric grid could produce efficiencies that would eliminate the need to build hundreds of new power plants. (***Source:*** L. D. Kannberg, D. P. Chassin, J. G. DeSteese, S. G. Hauser, M. C. Kintner-Meyer, R. G. Pratt, L. A. Schienbein, and W. M. Warwick, GridWise™: The benefits of a transformed energy system, Pacific Northwest National Laboratory, Richland, WA, PNNL-14396, 2004)

Local optimization of demand from buy-low, sell-high enforcement by RL controllers would transform usage patterns and supply/demand incentives at a micro level. Power management of congestion could be avoided by looping, multi-node, HVDC transmission lines connecting not only existing plants and substations but also terawatt-scale concentrations of solar power farms in deserts; giant wind farms on the plains; clean and safe nuclear, coal, stranded-gas, wave, and hydroelectric power plants; and offshore combined electric generation and carbon sequestration ships (i.e., FGSOs; see chap. 10).

It is vital to economic security that the electric grid be made more intelligent if we are to have any hope of preventing future disruptions of the electricity supply, as happened on August 14, 2003, in North America. In addition, the grid must evolve into the beyond-state-of-the-art system that the country will require for its next 50 years of economic growth, when the grid will be required to transmit and distribute 50% more power than today. Widespread use of far more advanced technologies will be needed in order to accommodate the additional power from massive, remote solar and wind farms. These sources are environmentally benign but erratic and unpredictable in their generation capacity. The present grid cannot cope with such variability.

A distributed-storage/generation electric grid will be adding interconnections to other nations as well. The European Union already has a plan for linking the electric grids of all of its member countries to an HVDC backbone. Thus, it will be all the more important for the electric economy of the future to cope with the added burden of defending our vulnerable populations and economic resources against infrastructure terrorism. Computer upgrades, along with new hardware (e.g., superconducting fault-current limiters, transformers and storage devices, digital power controllers, and next-generation nanotechnology transmission lines), will require testing and development into operational systems before they can be widely deployed.

The electric grid was originally interconnected to increase reliability and reduce cost. Interconnection means that the most expensive and environmentally dirty generators can be kept off-line if others—even a thousand kilometers away—can meet demand more effectively. However, without being smart, the transmission grid propagates fluctuations around the grid nonlinearly, and cascades can knock out transformers and substations and take generators off-line, as experienced in the Great Blackout of the Northeastern United States on August 14, 2003. In addition, operators must often limit a line's load, in some cases to as little as 60% of its ultimate thermal capacity, because transmission sometimes becomes unstable on its own, given its complex interactions within the grid. Power electronics combined with an intelligent control system can reclaim much of this lost capacity by using programmable processors that buffer surges and sags within a fraction of a second.

In a nutshell, the next-generation electric grid should be energetically efficient, economically competitive, environmentally friendly, technologically advanced, safe, and secure enough to transmit the massive amounts of energy that will be required to fuel our coming global electric economy. The following nanotechnologies will likely be important contributors if we are to achieve that vision of an electric economy by 2050:

- High-capacity, light-but-strong nano-cables using quantum conductors will rewire the transmission grid.
- Superbatteries and supercapacitors will allow massive distributed- storage/generation applications.
- Nano-electronics will revolutionize sensors and power control devices.
- Lightweight nano-materials will be used for high-pressure storage tanks, flywheels, and fuel cells.
- Distributed-generation fuel cells will drop in cost 10- to 100-fold.
- Nanotechnology lighting will replace incandescent and fluorescent lights.
- Nano-photovoltaics will drop in cost 10- to 100-fold.

PRICE SIGNALS

Electric utilities will be required to explore hundreds of possible configurations and outcomes to find the optimal systemwide solutions to customer needs as system capacity increases over the next 20 years. Connectivity to customers and their needs is key to success, yet most utilities have only rudimentary market understanding, much less customer-specific, sense-and-respond tools and techniques.

For modernization of the electric grid, the following new computer tools will be needed:

- Intelligent RL controllers for customer devices. The controllers should be able to eliminate arbitrage through the development of efficient energy markets that are scalable for the benefit of the consumer.

- Evaluation tools that continually meet the financial needs of customers in the new intelligent electric grid environment.
- A distributed ISM platform to compute optimal implementation strategies to match growing consumer demand with limited electric supply.
- Each asset of the system will contribute to its own simulation to arrive at optimal efficiency—a huge challenge.
- A systems architecture on such a scale would present many challenges to today's software engineering community. To create such an optimal system, CALM will be used for
  - Identification of the business capability and enumeration of objectives needed to define the required performance improvements.
  - Creation of road maps, chained matrices, and simulation models of the processes and workflows necessary to create the desired business capability.
  - Tracking of the actions that affect the processes of the business capability.
  - Quantification of the response of the system to those actions.
  - Identification of the locations of flexibility in the system.
  - Real options to utilize that flexibility.
  - Continuous reassessment of internal and external risks and uncertainties and mitigation of these risks through improvements in business capability.
  - Automated generation of steering signals at all levels of the operation, to drive the system toward more and more positive performance of the electric grid.

The CALM objectives are to map processes and create steering signals at multiple levels throughout the enterprise. Symptoms of trouble need to be detected and identified with problems, which then are mapped to solutions. However, metrics must then be created and efficiently captured to score the real performance of these solutions so that feedback can correct solutions that don't work so well. In the sections that follow, we give examples of how we anticipate that CALM will affect the coming electric economy, by examining first the vision for an energy-smart apartment house and then the added burden to the electric grid from the plug-in electric vehicle (PEV).

The energy-smart apartment house

Development of intelligent controllers for energy-smart apartment houses—and their interoperability with the outside world—requires electric load monitoring and control of appliances, such as refrigerators and air conditioners, that consume large quantities of electricity. Autonomous appliance controllers can shift the peaks in energy consumption by reacting to energy costs structured to be higher when demand is high. Verification and authentication of appliance availability from customers would then allow modulation of refrigerator and air conditioner usage to meet both the consumer's comfort levels and electric system constraints via pricing signals that drive more efficient use of the grid. The delivery of energy prices to a load-management gateway (LMG) in the basement on a real-time, hourly, and day-ahead basis could be easily attained by Internet, cable, and even bandwidth-over-power-line (BPL) communications. Load curtailment in response to price signals, coupled with ML analysis of customers' desired comfort patterns could unquestionably lower demand for electricity apartment by apartment, house by house, building by building, and city by city. For example, the curtailment of usage under high-demand conditions could be supported by intelligent appliances that will be able to negotiate with each other. They would jointly decide when to rotate turn-off and turn-on so that the electric consumption of the building is minimized.

In many areas of the United States, peak electric demand occurs in the summer, because of the use of air-conditioning. In major consumption states like New York, California, Texas, and Florida, 70% of the electric load on a hot summer day comes from a combination of air-conditioning and refrigeration. There are approximately 2.6 million window air conditioners in New York City alone. A utility can lower peak load demand by providing customers with the opportunity to save electricity without losing comfort. A *market-maker* in the smart controller of each appliance can intelligently decide to reduce customer's load on the basis of the energy price and customer preferences. When the load cycle allows it or the apartment needs it, as determined by price signals or temperature, respectively, the controller enables the appliance to turn on and report back to the controller that it is operating normally again.

Table 11–1 presents a *Merlin exercise* for the design of energy-smart apartment houses. In a *Merlin exercise,* the desired future state is envisaged, and then the steps and actions that would have to be taken to get there are listed by thinking back in time from the future to the present.

Table 11–1. Merlin exercise for the development of a demonstration pilot to develop autonomous appliances, controllers, and programs to create electric demand response in an energy-smart apartment house. AMI: advanced meter infrastructure.

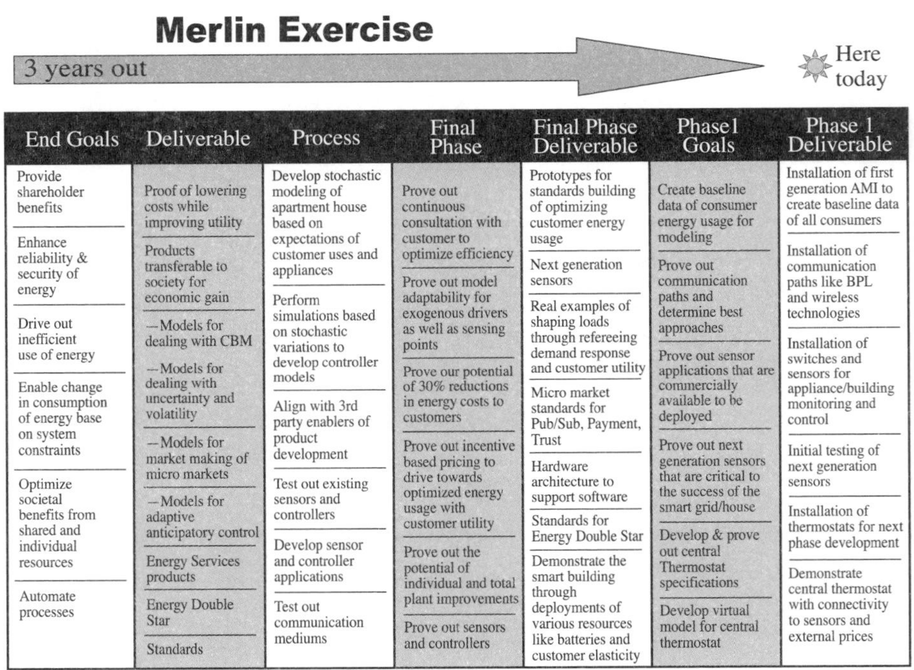

End Goals	Deliverable	Process	Final Phase	Final Phase Deliverable	Phase1 Goals	Phase 1 Deliverable
Provide shareholder benefits	Proof of lowering costs while improving utility	Develop stochastic modeling of apartment house based on expectations of customer uses and appliances	Prove out continuous consultation with customer to optimize efficiency	Prototypes for standards building of optimizing customer energy usage	Create baseline data of consumer energy usage for modeling	Installation of first generation AMI to create baseline data of all consumers
Enhance reliability & security of energy	Products transferable to society for economic gain	Perform simulations based on stochastic variations to develop controller models	Prove out model adaptability for exogenous drivers as well as sensing points	Next generation sensors	Prove out communication paths and determine best approaches	Installation of communication paths like BPL and wireless technologies
Drive out inefficient use of energy	—Models for dealing with CBM	Align with 3rd party enablers of product development	Prove our potential of 30% reductions in energy costs to customers	Real examples of shaping loads through refereeing demand response and customer utility	Prove out sensor applications that are commercially available to be deployed	Installation of switches and sensors for appliance/building monitoring and control
Enable change in consumption of energy base on system constraints	—Models for dealing with uncertainty and volatility	Test out existing sensors and controllers	Prove out incentive based pricing to drive towards optimized energy usage with customer utility	Micro market standards for Pub/Sub, Payment, Trust	Prove out next generation sensors that are critical to the success of the smart grid/house	Initial testing of next generation sensors
Optimize societal benefits from shared and individual resources	—Models for market making of micro markets	Develop sensor and controller applications	Prove out the potential of individual and total plant improvements	Hardware architecture to support software	Develop & prove out central Thermostat specifications	Installation of thermostats for next phase development
Automate processes	—Models for adaptive anticipatory control	Test out communication mediums	Prove out sensors and controllers	Standards for Energy Double Star	Develop virtual model for central thermostat	Demonstrate central thermostat with connectivity to sensors and external prices
	Energy Services products			Demonstrate the smart building through deployments of various resources like batteries and customer elasticity		
	Energy Double Star					
	Standards					

In the desired future state, an energy-smart apartment house can be made airtight except on beautiful days, with insulated construction to reduce heating and cooling requirements. It has a green roof, a rooftop garden with sufficient plant area to offset CO_2 generation elsewhere on the premises. The green roof can reduce building temperatures by as much as 10°F and hold enough photovoltaics and windmills to offset 10% of annual energy needs. There are low-flow plumbing fixtures in every apartment. To the extent possible, recycled rainwater flushes toilets and is used on the roof for irrigation. Heat-pump chillers produce both high-efficiency air-conditioning in the summer and enough recovered heat to warm apartments in the winter.

Every apartment has an intelligent motion sensor to detect movement and turn off all lights when not needed. Every apartment has what we call *energy double star appliances,* where smart RL controllers modulate the energy usage, determined by the consumer's preferences to save energy and live in comfort. Such *green and efficient* apartment houses would cut electricity use by 40% and water use by 30% each, compared to traditional buildings.

To be sure, much success has been achieved with programs like Energy Star, which labels efficient appliances in the United States, but governments can do much more to promote energy efficiency and intelligently curtail energy usage. It comes down to providing trusted and measurable benchmarks and incentives for consumers to make intelligent decisions. The same can be done with all products and services, by providing a sustainability index, a metric that quantifies the degree to which the overall home is environmentally sustainable as compared to others nearby. Bonuses can be paid to the most efficient homes by the utilities. In addition, standardization of communications, coupled with promotion of autonomous appliance modulation, is best achieved through government collaboration with utilities and energy services companies.

We have seen an industry-wide move toward high-technology metering for customers, referred to as *advanced meter infrastructure* (AMI). Figure 11–6 lists potential benefits a utility may consider in its AMI build-out that will make the electric grid smarter. Most electric utilities are striving for more than automation of meter reading, which at present primarily entails employees walking door to door to read dials on meters. However, there are underlying concerns with these deployments. Today, meter reading has an average cost of about $3.00 per meter per month. New York City has three million electric meters.

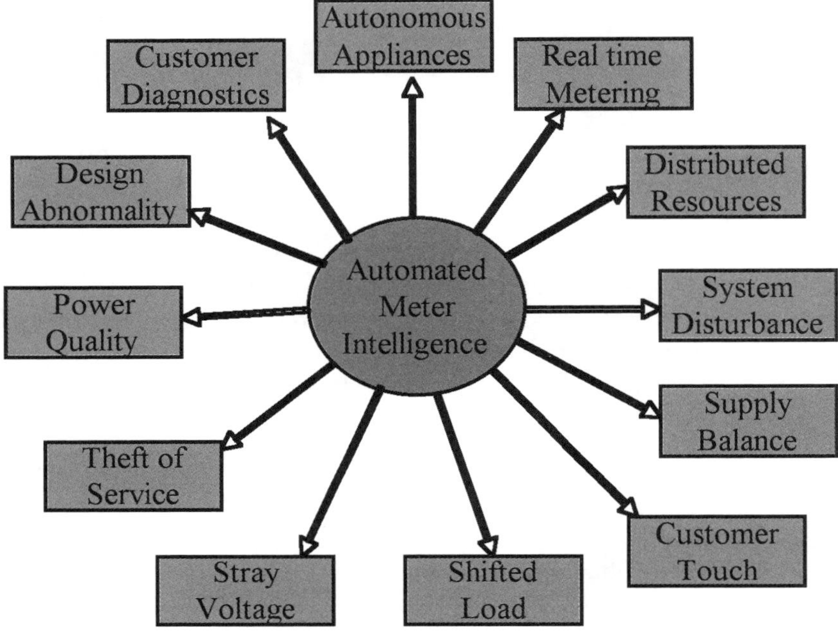

Fig. 11–6. Potential benefits from AMI to enable the intelligent electric grid

Where is the plan to make sense of the terabytes of additional data that will be coming with AMI so that energy-smart customers can be enabled? Where are the resources committed to making intelligent decisions that can design the right price signals to send and other incentives? Through commercially available software packages, accurate research statistics on consumer usage are readily available to extract a wealth of information about consumption that has already led to better comprehension of the state of the electric grid. However, most utilities have not fully used the data coming in from AMI.

Many consider AMI to be the enabling technology for the intelligent grid of the future, yet there is little or no new technology in most AMI meters manufactured today. The following are examples of the value from the deployment of even smarter AMI:

- Customer touch. Currently, only your cable television box knows when a customer has an electric outage in his home. AMI extends this real-time sensing capability to the utilities. However, there is no improved customer satisfaction if this does not result in the return of better information to the customer and faster recovery operations should the lights go out.
- Design abnormalities. Today, engineers must analyze billions of data points in a particular area to detect potential performance abnormalities—and each time on a one-off basis. AMI has the potential to automatically detect costly operating variances, like overloaded transformers, that could harm the public or workers. However, smart computational algorithms would have to be deployed in order to identify whether such were from design abnormalities.

The preceding examples support the need for large additional investments in computer intelligence to go along with the massive commitments for the physical deployment of AMI that are already planned. Following are the approximate costs to some representative U.S. electric utilities for deployment of AMI:

- CenterPoint, Houston, TX: $1,800 million
- Pacific Gas & Electric, Northern and Central CA: $1,700 million
- Southern California Edison, CA: $1,300 million
- San Diego Gas & Electric, CA: $600 million
- Con Edison, New York, NY: $900 million
- Baltimore Gas and Electric, MD: $400 million
- Southern Company, Southeastern United States, GA: $280 million

On top of the close to $7 billion in AMI costs, investment is still required in order to develop the business intelligence capabilities in order to leverage this information to extract maximum value to customers from this flood of new data. Unfortunately, intelligence development is not part of any of the preceding AMI deployment budgets. CALM technologies have made sense from such information overloads in other industries and can do so for AMI data as well—and at a cost of 10% or less of the preceding deployment budgets.

Plug-in vehicles

The greatest human-generated danger to Earth's global climate comes from cars and trucks. Automobiles produce more CO_2 than all other greenhouse gas–emitting sources combined. In addition, they account for 60% of the hydrocarbon consumption on Earth.

Conversion from internal combustion engines to electric vehicles must overcome an inherent problem, however. Internal combustion engines allow extremely cheap and efficient transportation; while the engine itself is not energy efficient, the vehicle is. Conversion to electric vehicles has not been practical so far because of battery life, but that might be about to change. The first electric taxis were promoted in New York City by Thomas Edison at the turn of the 20th century. Henry Ford quickly bumped them off the roads with the Model T. (Incidentally, Edison and Ford were the best of friends.) Fast-forward to the present. Electric vehicles are making a comeback, buoyed by new battery and engine technologies, environmental concerns, the high cost of gasoline and diesel, and dependencies on monopolistic sources of oil from OPEC and the Middle East.

A combination of the internal combustion engine and an electric motor is the dominant innovation at the beginning of the 21st century. These *hybrid* vehicles take advantage of the benefits of both electricity (rapid and efficient acceleration and deceleration) and gasoline (efficient cruising). A version of the hybrid vehicle called the *plug-in hybrid electric vehicle* (PHEV) uses electricity from a combination of sources. PHEVs are able to tap electricity stored in batteries that are charged from plugging into the electric grid when the vehicle is not in use. They can also generate electricity on board when the internal combustion engine is in use and from braking energy.

There are currently two types of PHEVs in development: (1) hybrids with parallel engines (e.g., the Toyota Prius), which use a gasoline engine that powers the wheels through a transmission and batteries that drive the wheels through an inverter that in turn powers a motor; and (2) hybrids with series engines (e.g., the Chevrolet Volt), in which batteries and the internal combustion engine both produce electricity to drive a generator that runs the motor that turns the wheels, all in series. With either design, the batteries and the internal combustion engine can be used anytime. An onboard computer decides what mix is most efficient. In general, the cars use electricity from a stop to cruising speed, and then they switch to the gasoline engine, which is more efficient at steadier speeds. This combination triples gas mileage.

A PHEV also has the ability to recharge its batteries from the electric grid when not in use. Plugging the vehicle into the electric grid extends its all-electric range to at least 40 miles before the gas engine would need to be turned on. If all future trucks and cars on U.S. roads were PHEVs, that would cut dependence on imported oil in half. More than 60% of all trips that Americans take are less than 40 miles and therefore might require no gasoline consumption at all. A switch to plug-in electric vehicles (PEVs) seems to be the natural progression of this technology. When this transition occurs, the world will truly enter the era of the electric economy.

In addition, PHEVs and PEVs present enormous environmental benefits. An internal combustion car that gets 25 miles per gallon emits one pound of CO_2 per mile, or 25 pounds per gallon. Today's hybrid vehicle gets 50 miles per gallon and emits 0.65 pounds per mile, or 32 pounds per gallon. A PEV does much better. The vehicle itself emits no CO_2, but it uses electricity from power plants. The worst coal-fired power plant in the United States today emits 0.7 pounds of CO_2 per gallon equivalent of fuel that it burns (measured in British thermal units), which is an order of magnitude better than any gasoline car or diesel truck now, hybrid or otherwise. The average for the whole U.S. electric grid is 0.5 pounds, and states that require at least a 15% renewable resource mix (mostly from hydroelectric power plants) bring that average down to less than 0.4 pounds; further, nuclear power plants emit zero pounds (but of course have the high-level radioactive-waste disposal problem) (fig. 11–7).

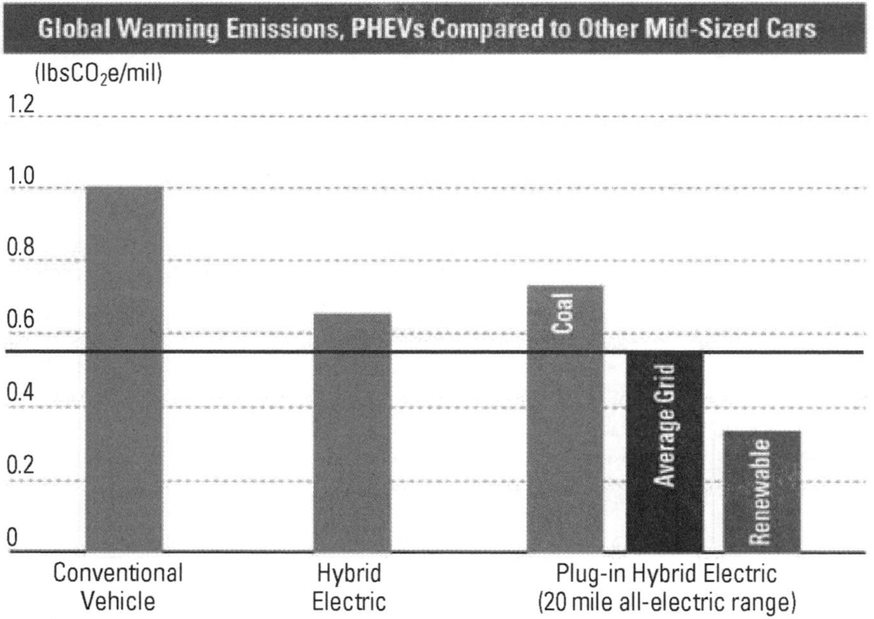

Fig. 11–7. Environmental benefits of hybrid vehicles (***Source:*** EPRI-NRDC Joint Technical Report on Plug-In Hybrid Electric Vehicles, July 2007)

Currently, cars and trucks use more than half of all the oil imported into the United States, cause 90% of smog emissions even with catalytic converters, and account for 25% of all greenhouse gas emissions. Hybrid cars use electricity that costs the equivalent of up to one dollar per gallon of gasoline. Hybrid vehicles also receive benefits from states. For example, California allows them unrestricted access to high-occupancy vehicle (HOV) lanes and takes money from insurance paid for gas-guzzling sports-utility vehicles (SUVs) to subsidize cheaper rates for hybrid owners. In the near future, hybrids can expect to receive carbon emission credits and avoid congestion pricing in high-traffic areas. Electric vehicles should receive even more benefits.

So what market challenges are there to making first hybrid and then electric vehicles the norm? The principal challenge is battery technology. Battery life, reliability, safety, weight, recharge speed, purchase and maintenance cost, and environmental footprint must all be improved before hybrid and electric vehicles can become economical. The PEV battery faces an added burden of withstanding deep discharges and daily cycling. PHEVs now are very expensive, with a differential cost estimate of $18,000 over gasoline vehicles, because additional batteries and controls are required.

Recent marketing surveys indicate that while hybrids represent only 2% of new car sales, 50% of new buyers said they would consider a hybrid for their next car if it were available in the class of vehicle they use, and 26% would pay a $4,000 premium.[5] Governments, utilities, and automobile manufacturers would all have to integrate their efforts to make PHEVs—let alone PEVs—fit the aforementioned consumer-buying pattern.

Transportation load added to the electric grid

The current electric grid already supplies power for significant urban transportation, such as subway and commuter rail systems. If PHEVs, then PEVs, become the norm, the grid will be taxed beyond present levels. The worst-case scenario involves charging as soon as commuters get home at night and then again when they get to work. Con Edison and New York City have even investigated making all-new parking "meters" that have power plugs. Not only would this double the load growth in all urban metropolises, but it would increase peak demand in the mornings in commercial districts and in the evenings in residential areas by 25% (fig. 11–8).

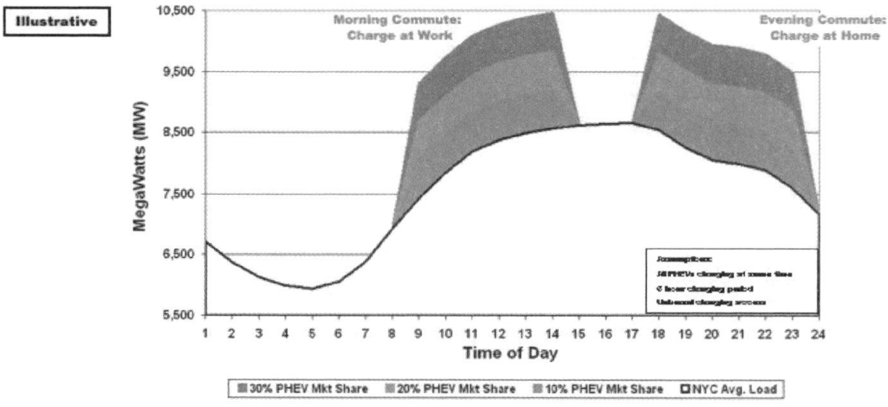

Fig. 11–8. The effect of PHEVs on average weekday electricity load cycle (***Source:*** Con Edison)

A best-case electricity consumption pattern would add to each hybrid vehicle smart chargers that receive price incentives to charge only during low-demand times (e.g., after the morning rush hours and late at night) (fig. 11–9). In addition, their batteries would have the potential to supply electricity back to the electric grid in times of system emergencies. The option of directing a modest supply of electricity from PHEVs and PEVs into the grid would require potential increases in battery size and changes in inverters and controllers to allow flow out of the vehicle.

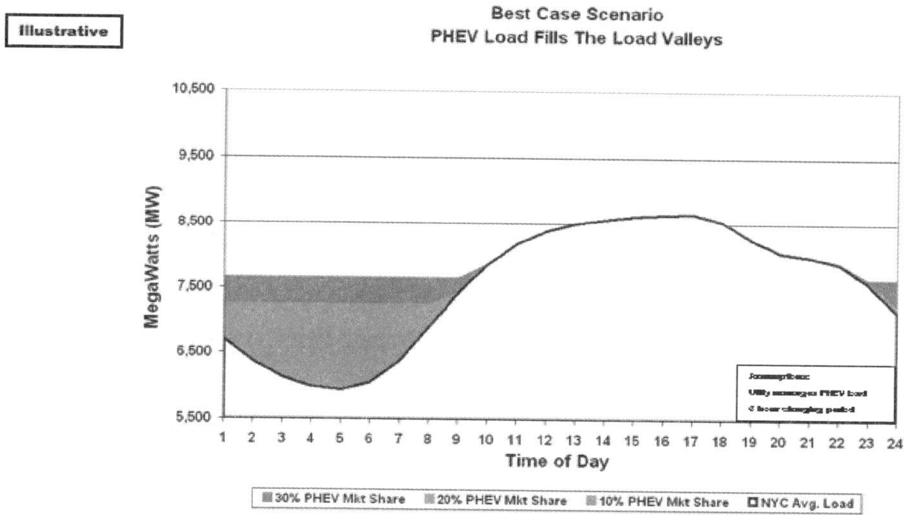

Fig. 11–9. The effect of using smart chargers for PHEVs on the average electricity load cycle. Managing demand for vehicle charging will require close coordination between utilities and automakers, to prevent the grid from being overwhelmed. (***Source:*** Con Edison)

Economic incentives—through time-of-day rates, coupled with intelligent controllers—could make off-peak vehicle recharge a reality. This change brings its own set of unknowns. Current peak-to-trough electric usage cycles in urban areas allow the copper wires to cool off at night. The pattern driven by heaver nighttime consumption would change this heating and cooling cycle. This may be advantageous—or not—to the reliability of the electric grid. We do not yet know, but the ISM will offer the simulation capability to model these various futuristic scenarios.

In any event, city planning, transit agency, utility, and automotive industry collaborations are necessary before such large and unpredictable loads can be safely added to the electric grid. The immediate problem would occur on the distribution grid. It would have to become much smarter and more like the Internet to cope with this added load, as discussed previously. The grid will need to reroute electricity to cope with dramatic geographic shifts in consumer demand if PHEVs and PEVs are added to subways and commuter rail systems as electric power becomes the dominant form of transportation. Because of the benefits from carbon reduction, fuel independence, and reduction in the use of imported oil, there are strong incentives for all parties to unite to work toward such a transportation future built on the electric economy.

The first PHEV is not projected to enter the marketplace until 2010, and the likely adoption profile predicts substantive impact on daily load curves for all urban utilities by 2020. That adoption rate assumes the right prices and incentives occur along with the likely breakthroughs in battery technology. For example, if the PHEV has peak demand capabilities that can be curtailed, additional capacity requirements of generation are not needed, enabling the potential for reduced costs for electricity. The challenge will be to create the electric economy that is driven by an intelligent grid that is evermore controllable at ever-finer scales. That will require the processing of orders of magnitude more data than were coming in from turn-of-the-21st-century SCADA systems, which will seem primitive at best by then. Terabytes of real-time data will need to be processed in real time using entirely new methods of integrating information and models. This will require the CALM methodology to interpret the real-time data streams by using advanced ML and pattern recognition to turn data variances into actionable information, much like spy satellites do now.

Otherwise, blackouts (as in the Northeastern United States in 2003) and brownouts (as in California in 2000) could shut down not just cities and subways but also automobile and truck travel. Grid disruptions would become even more unacceptable than they are now. Utilities cannot be insular in such a future world; neither can oil companies or automakers; and, with the potential of peak oil coming, neither can politicians, city and state administrators, or the citizenry. We will all be huge stakeholders in the electric economy.

INTELLIGENT CONTROLLERS

When the electric grid is asked to perform at higher levels, equipment failures occur more frequently, posing safety risks to the public and workers. Higher rates of capital expenditures and operating costs are then required. Since initial failures often introduce destructive transients to ancillary and neighboring equipment, resulting in additional failures that can cascade into severe crises, it is important to reduce the frequency of these emergency failures, to every extent possible.

The future intelligent electric grid will likely have a control system that analyzes its performance using distributed, autonomous RL controllers that have learned successful strategies to govern their behavior in the face of an ever-changing environment (fig. 11–10). Such a system might be used to control electronic switches on primary distribution feeders that are tied to multiple area substations with varying costs of generation and reliability. The R&D to achieve such a self-adaptive optimizing grid has several convergent fronts, described in the following section.

Learn from Nature --
Adaptive Control Works at *All* Levels

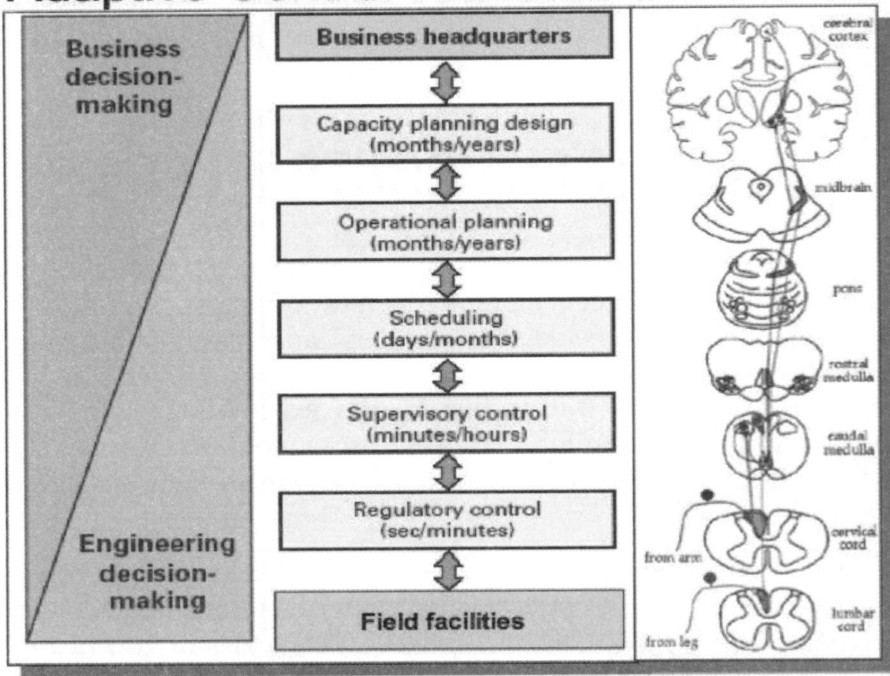

Fig. 11–10. Integrated sets of monitoring and control systems. Cooperation among these systems will need to be maintained in order to reduce the frequency and duration of excursions into operating conditions under which the system is more failure prone and therefore less efficient in delivering services to customers. Nature provides such control through the central nervous system. (***Sources:*** Left and center: L. A. Saputelli, S. Mochizuki, L. Hutchins, T. Cramer, M. B. Anderson, J. B. Mueller, A. Escorcia, et al., Promoting real-time optimization of hydrocarbon producing systems, Society of Petroleum Engineers paper SPE 83978, 2003; right: Somatosensory pathways from the body, The Washington University [St. Louis] School of Medicine [http://thalamus.wustl.edu/course/body.html])

RL components

How do we achieve this smartening of the grid? The CALM approach in the future will be to use reinforcement learning (RL) controllers as a basis for optimizing the synergy between operators and automation. Luckily, cheap silicon has been invented that is small enough to attach to critical equipment, identify itself, geo-locate using GPS, and contain as much memory and processor power as a laptop computer. This silicon will wirelessly communicate with central control over secure ultra-wideband Wi-Fi, all the while incorporating sensors via micro-electro-mechanical (MEM) chips that communicate with other sensors. The sensors, computer, and communicator will be self-contained and operational for the total life cycle of the asset, from creation to destruction. We call this the *silicon life cycle.*

Addition of real-time sensing and control is crucial if the future grid is to exploit synergies between variable (wind and solar) power sources and dispersed energy storage solutions that can buffer the transmission and distribution system. The key to exploiting these synergies to their fullest potential is ubiquitous silicon associated with all critical assets in the electric grid—from generation to storage, transmission, distribution, and consumption.

We must grow the simulation capabilities to encompass design of work layouts, hardware options, and costs for other power supply buffers such as superconducting fault-current limiters, transformers, and distributed-generation and -storage devices (e.g., fuel cells and flywheels). All will need to be tested on the computer before physical designs are finalized.

The RL controller will need to be adaptive to continuously changing operating conditions. It will therefore also need to tolerate uncertainty and optimize control actions on the basis of real-time inputs. This stochastic, adaptive, RL controller hierarchy will need to integrate tightly with financial decision-making functions by using operational objectives based on risk, reliability, environmental impact, efficiency, and safety.

INFRASTRUCTURE INTERDEPENDENCY

Among the great human accomplishments is the vertical city, with electricity powering the elevators and water pumps that make skyscrapers possible, as well as the urban rail and subway services that transport the population efficiently and quietly. In addition, electricity has enabled the conquering of diseases like tuberculosis, the purification of drinking water, and the alleviation of summer heat and winter cold. It has ushered in the digital age of information, telecommunications, and the Internet.

Yet, people don't buy electricity. Most people hardly know how much it costs, and they take for granted that it will be available every time they flip that light switch or turn on the television. Instead, they value the information, entertainment, security, light, and comfort that electricity brings. They are unaware of the difficulties in keeping the supply of electricity available 24/7. It falls to Con Edison (in New York), Tokyo Electric and Power, Électricité de France (in Paris and London), and the other great urban utilities of the world to keep these vertical cities powered so that they can remain safe, clean, and civil.

The electric economy would force the electric grids of modern cities to be reconfigured over the next 10–20 years in other ways beyond intelligence and efficiency. The potential effects of upgrading and modernizing the electric grid would touch every aspect of commercial, industrial, and consumer activity in the cities of the world. The electric grid would be only a part of the intelligent grid that cities require. That extended grid will likely also include integration with other key infrastructures in cities, such as water, sewage, telecommunications, heating oil, natural gas, and electric transportation. The great cities will require an ISM of the combined infrastructure systems as well (fig. 11–11).

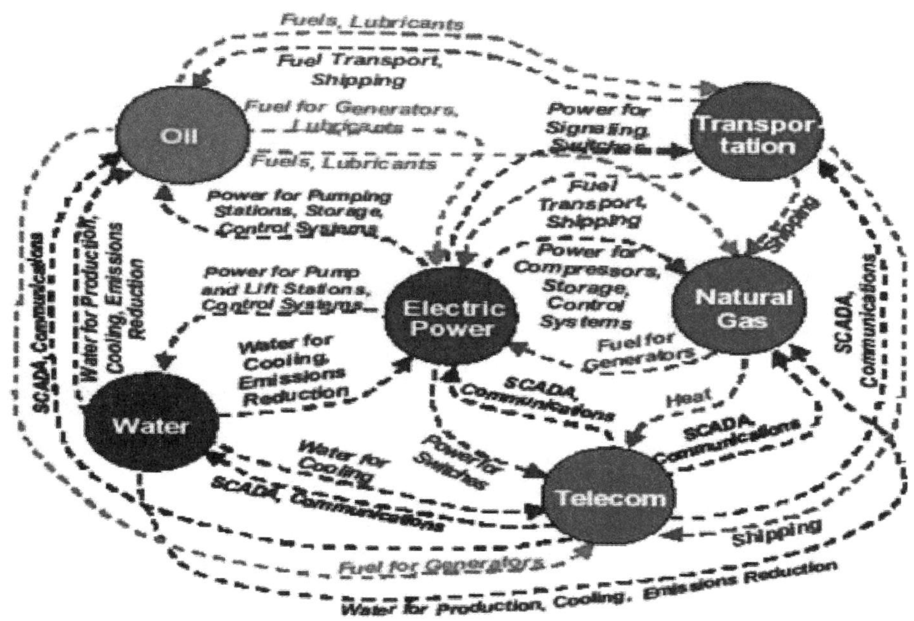

Fig. 11–11. Complex interdependencies between infrastructures—a challenge for the integrated-infrastructure intelligent grid (***Source:*** J. Peerenboom, Infrastructure Assurance Center, Decision and Information Sciences Division, Argonne National Laboratory, 2001).

We envisage that the ISM that controls this intelligent electric grid will have real-time RL systems attached. The combination will allow the creation of virtual SCADA predictions for infrastructures that are not sufficiently monitored at present, such as the water system in the United States. Major cities in Canada and Europe, such as Toronto and Paris, use a SCADA system to monitor water delivery (e.g., Water Security Management Assessment, Research, and Technology [W-SMART] [http://uucl.poly.edu/wsmart]).

Once the electric, gas, water, and rail systems are combined into one ISM, the RL controller will be able to simultaneously solve for the most operationally and economically viable solutions and outcomes among hundreds of possible configurations and scenarios for future modernization of the combined infrastructures of the city. For example, we envisage the creation of an ISM for the combined electrical systems of Con Edison and the rail systems of the Metropolitan Transit Authority (MTA) in New York City (fig. 11–12).

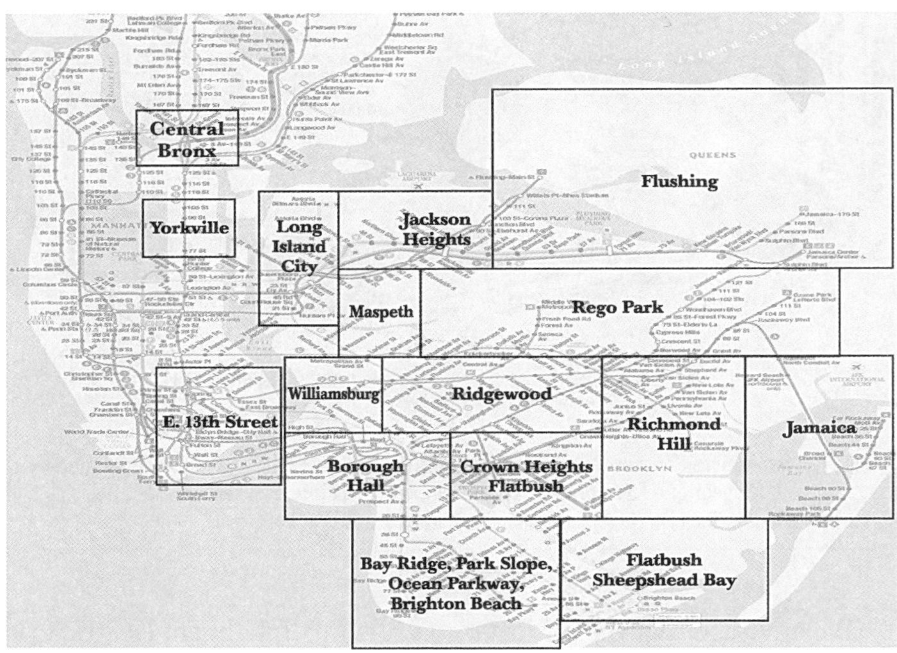

Fig. 11–12. An electrical backbone running along subway and rail tunnels, such as in New York City (above), with distributed generation and storage at key subway and electrical substations (using superbatteries, fuel cells, wind, solar, etc.). Dramatic improvements in the reliability of subway power, coupled with multi-megawatt load-shifting capabilities for the utility, might be accomplished in order to offer significant improvements in resiliency, reliability, and efficiency of both the subway system and electric grid.

New York (and other cities) can use added resiliency, sustainability, and security to weather the storms and other emergencies its citywide electric and rail systems encounter. Power delivery congestion can be alleviated by passing power through the third rail or along specially laid electrical feeder cables running through rail and subway tunnels. Electric load balancing would also have to coordinate with power generation within the urban grid. Costs and benefits could be evaluated using the ISM to investigate more collaboration within integrated systems. Steel third rails that carry the electricity might be replaced in older subway systems with electrically conducting aluminum, as in the new Dallas Area Rapid Transit (DART) system. The added resiliency to weather and other emergencies may well make the conversion cost-effective; scenario analysis would tell.

The intelligence and control software needed for combined operation of the electric, gas, steam, water, telecommunications, and rail infrastructures have yet to be built. These infrastructures have in common centralized control-centers. We foresee the development and simulation of an integrated control environment for cities that resembles the battlefield visibility systems of the U.S. Army and damage control systems in U.S. Navy ships. Major questions include how much operating data must be shared among control-centers for each infrastructure to optimize its performance while assisting others. Connectivity to customers and attention to their needs are also keys to successful optimization, yet most infrastructure control-centers now have only rudimentary customer-specific, two-way sense-and-respond tools and techniques (fig. 11–13).

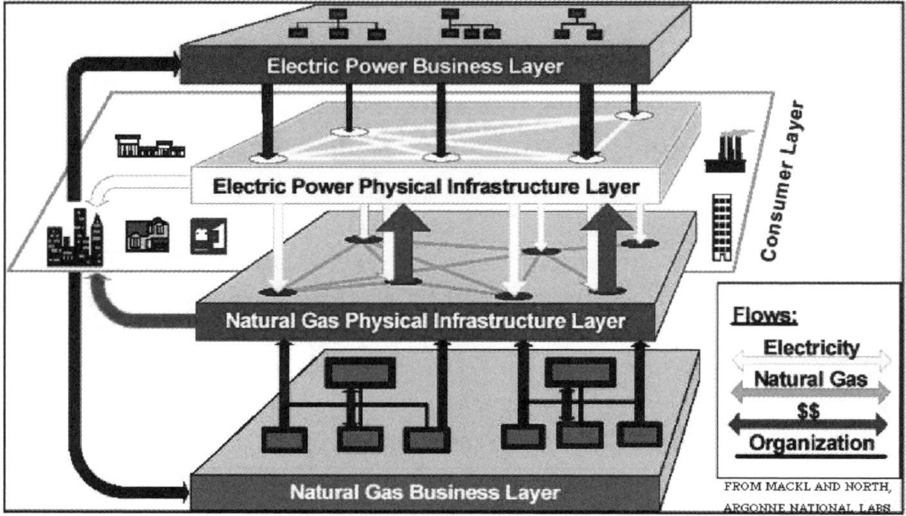

Fig. 11–13. Layers of interdependencies of electric and natural gas systems. These will become more interrelated and complex in the future. (***Source:*** C. Macal and M. North, Electricity Market Complex Adaptive System, Argonne National Laboratory [http://www.dis.anl.gov/projects/emcas.html])

In the future, CALM will strive to provide additional core competencies in innovation and technology to the intelligent grid to assist in the development of improved performance of the electric, gas, steam, water, rail, and subway grids. That expertise supersedes the engineering and construction capabilities that infrastructure engineers are familiar with today. The next generation of integrated-systems engineers will need business as well as engineering expertise. The focus will be on transforming the use of technology to replace emergencies with scheduled remediation and to enable true integrated-systems management. The successful integration of all critical infrastructure will provide the resiliency that cities will need in the future.

FUTURE ELECTRIC ECONOMY

As we have said, the technical focus of CALM within the future electric economy will likely be on the development of the necessary computer intelligence. The intelligent electric grid will require sustainability, reliability, resiliency, and efficiency. A smart, RL controller for the ISM will hopefully enable the intelligent electric grid to operate.

The ISM is needed so that the integration of the electric grid can be designed and tested on the computer first, with consequences evaluated before actions are taken. The ISM then becomes the intelligent master simulation for the distributed field systems that consumers use, like PHEVs, local distributed generation from renewable resources in the area, and power storage. The aim of the modeling effort will be to incorporate the ability to optimize across the many different infrastructures that require distributed controllers, concentrating the feedback from the field systems into an RL controller that determines optimal policies for operating the integrated infrastructures of the intelligent electric grid. The scope is illustrated in this vision of the family home in 2020 (fig. 11–14).

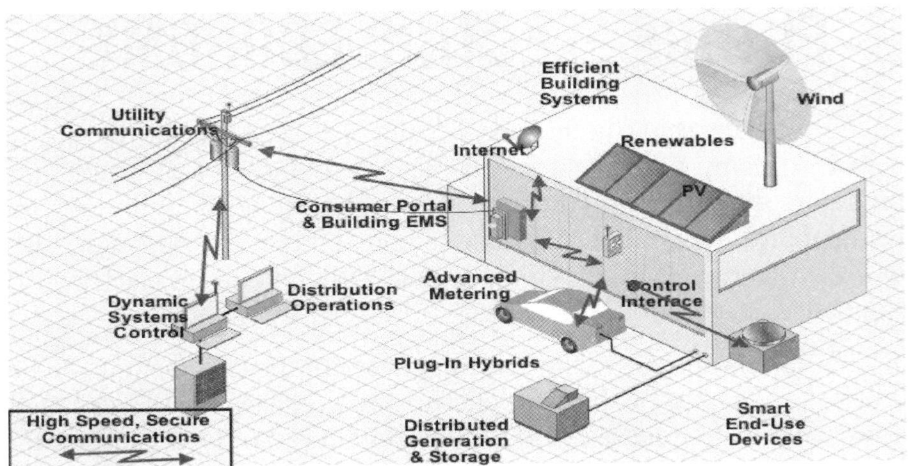

Fig. 11–14. Family home of the future. A simulation environment is needed to determine how to run the family home most efficiently in this vision of what it might look like in 2020.

With the feedback loop from RL, the ISM will become the glue that holds the design models, control systems, decision support, operations, and maintenance together with the consumer, price signals, and the market. This integration will eventually make the grid truly intelligent (fig. 11–15).

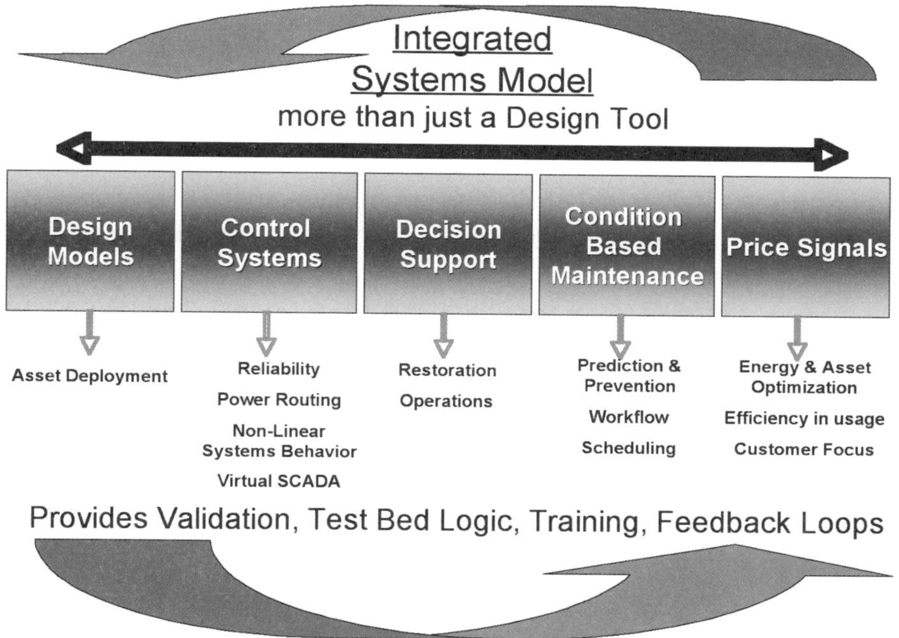

Fig. 11–15. ISM of the smart grid. The ISM provides the glue that will holds the intelligent electric grid together.

All aspects of the intelligent grid will be affected—from the creation of new preventive maintenance software programs and the consolidation of operations software to the modernization of field repair, to finance, and to the interaction with customers and regulators alike. Not only will transmission and distribution be changed, but supply chains, operations, maintenance, and energy services will all be profoundly reconfigured. These changes will need to be accepted and understood by the new energy industry workforce.

An intelligent grid is necessary to provide optimal, efficient, clean, plentiful, safe, and cheap energy to power continued economic development. We have the chance to create a superefficient intelligent grid with enough elastic demand response to require only a limited number of new fossil fuel power plants and allow a maximum number of giant solar and wind farms.

We can envisage the technologies to create such an intelligent grid that will be economically competitive, technologically advanced, safe, and secure, as well as able to transmit massive amounts of electricity (by today's standards) over large distances. How to get there from here is a much harder question. Although significant technical problems remain—for example, how best to connect distributed generation and distributed storage to the present grid. The lack of interconnection capabilities of many conventional distributed-generation devices to the grid has been identified as one impediment to their use The need for better interconnection standards, not technology upgrades, is the most glaring problem at present. By *interconnection standards,* we mean out-of-the-box plug-and-play, self-realization, and capability to connect to the grid without the potential to do harm. Presently, the costs of many interconnections make them uneconomical. Also, there are no incentives for fixing the grid, beyond short-term patches (e.g., laying additional dumb transmission wires around congestion). That strategy echoes urban highway construction: the more lanes provided, the more traffic the road attracts, producing more congestion, requiring more lanes, and so forth.

For the intelligent grid to be fully successful, national test beds are needed, to determine how to deploy the new intelligent grid technologies on a massive scale. Test beds would combine promising technologies and experiment with how the system is improved through their use (difficult to do when individual technologies are deployed in isolation). For example, control software might be deployed to manage power from zero-emission coal gasification power plants and offshore wind farms, linked by the nation's first quantum wire transmission system and connected to a combined AC/DC urban distribution system with abundant local storage and substation routing technologies.

Such test beds should already be in operation if we are to meet the power needs of large cities 20 years from now. It is difficult, costly, and time consuming to experiment with the live grid that provides service to customers. We must be certain that the grid is capable of handling each new technology before it is deployed. It is not an option to connect new gadgets directly to the grid and accidentally cause harm or massive, cascading blackouts. The problem with creating such test beds is that the electricity industry has among the lowest R&D expenditures of all Wall Street industrials.

Coal, oil, and gas alone will not bring enough energy to the 22nd-century population. Yet, the electric grid is aging fast, along with other interrelated infrastructures, such as steam, gas, water, transportation, and sewage. We are running out of time, and CALM solutions like those we have discussed in this book are needed. To develop the intelligent grid to its utmost and extract the value from such a superior system, the energy industry and the government must significantly increase funding and resources in a collaborative fashion and use computer intelligence to build the capabilities needed to develop and implement the integrated-infrastructure grid of the electric economy.

NOTES

[1] Testimony of the late Rick Smalley, Nobel laureate and pioneer of the buckyball and the carbon nanotube, to the U.S. Senate Committee on Energy and Natural Resources at the April 27, 2004, hearing on sustainable, low-emission electricity generation.

[2] Energy Information Administration, U.S. Department of Energy. July 31, 2006. *World Consumption of Primary Energy by Energy Type and Selected Country Groups, 1980–2004.*

[3] America's electricity crisis. 2003. *The Economist,* August 23.

[4] The electric grid was named the greatest engineering achievement of the 20th century by the National Academy of Engineering.

[5] L.A. Auto Show 2007; Merrill Lynch, New York governor's press release. December 22, 2006, Forbes, NYPA.

FURTHER READING

Columbia University CALM publications

Anderson, R. 2000. Technical innovation: An E&P business perspective. SEG Leading Edge. June.

———. 2004. The distributed storage-generation "smart" electric grid of the future. In The 10-50 Solution: Technologies and Policies for a Low-Carbon Future. PEW Center for Global Climate Change and the National Energy Policy Council.

Anderson, R., and A. Boulanger. 2002. ThreatSim: Securing wattage when needed. Energy Pulse. October 14.

———. 2003a. Lean energy management I: Lean energy management required for economic ultra deepwater development. Oil & Gas Journal. March 17.

———. 2003b. Lean energy management II: Ultradeepwater development: Designing uncertainty into the enterprise. Oil & Gas Journal. May 19.

———. 2003c. Lean energy management III: How to realize LEM benefits in ultradeepwater oil and gas. Oil & Gas Journal. June 30.

———. 2003d. Lean energy management IV: Flexible manufacturing techniques make ultradeepwater attractive to independents. Oil & Gas Journal. August 25.

———. 2003e. Lean energy management V: Enterprise-wide systems integration needed in ultradeepwater operations. Oil & Gas Journal. November 24.

———. 2004a. Lean energy management VI: Ultradeep offshore suitability matrix for estimating value of lean processes. Oil & Gas Journal. June 28.

———. 2004b. Lean energy management VII: Knowledge management and computational learning for lean energy management. Oil & Gas Journal. November 24.

———. 2004c. Smart grids and the American way. Power & Energy, American Society of Mechanical Engineers. March.

———. 2005a. Lean energy management VIII: Use of matrices in computer-aided lean energy management. Oil & Gas Journal. March 7.

———. 2005b. Lean energy management IX: Boosting, support vector machines and reinforcement learning in lean energy management. Oil & Gas Journal. May 9.

———. 2005c. Lean energy management X: How martingale stochastic control navigates computer-aided lean energy management. Oil & Gas Journal. September 19.

———. 2007. Lean energy management XII: Real options of gas-to-electricity in ultra deepwater production. Oil & Gas Journal. July 23.

Anderson, R., A. Boulanger, W. He, U. Mello, and L. Xu. 2007. Lean energy management XI: 4D seismic reservoir monitoring with CALM's martingale jontroller. Oil & Gas Journal. March 31.

Anderson, R., A. Boulanger, J. Longbottom, and R. Oligney. 2003. Future natural gas supplies and the ultra deepwater Gulf of Mexico. Energy Pulse. http://www.energypulse.net/centers/article/article_display.cfm?a_id=232

Anderson, R., et al. 2001. Power shortage. CIO Insight. July.

Anderson, R., and W. Esser. 2001. Energy company as advanced digital enterprise. American Oil & Gas Reporter. January.

Esser, W., and R. Anderson. 2001. Visualization of the advanced digital enterprise. Paper OTC 13010, presented at the Offshore Technology Conference, Houston.

Gross, P., R. Anderson, et al. 2007. Predicting electricity distribution feeder failures using machine learning susceptibility analysis. International Association of Artificial Intelligence. July.

Machine learning

Alpaydin, E. 2004. Introduction to Machine Learning. Cambridge, Mass.: MIT Press.

Anderson, R., and A. Boulanger. Stochastic controller for real time business decision-making support. International Patent Cooperation Treaty PCT/US04/28185.

Anderson, R., A. Boulanger, J. Johnsohn, and A. Kressner. 2006. Getting lean and efficient. EnergyBiz Magazine. July/August.

Anderson, R., et al. 2006. Systems and methods for martingale boosting in machine learning. United States Letters Patent (applied for).

Anderson, R., et al. 2007. System and method for grading electricity distribution network feeder conditions. United States Letters Patent (applied for).

Anderson, R., et al. 2007. Propensity as a way of predicting survivability of electrical feeder components. Provisional United States Letters Patent (applied for).

Anderson, R., et al. 2007. Decision support system for electric control center operational optimization. Provisional United States Letters Patent (applied for).

Becker, H., and M. Arias. 2007. Real-time ranking of electrical feeders using expert advice. Paper presented at the European Workshop on Data Stream Analysis, Caserta, Italy. http://www.lsi.upc.edu/~marias/papers/wdsa.pdf

———. 2007. Real-time ranking with concept drift using expert advice. KDD. August. http://www.lsi.upc.edu/~marias/papers/kdd07.pdf

Bellman, R. E. 1957a. Dynamic Programming. Princeton, N.J.: Princeton University Press.

———. 1957b. A Markov decision process. Journal of Mathematical Mechanics. 6: 679–684.

Bishop, C. M. 2006. Pattern Recognition and Machine Learning. New York: Springer.

Brown, J. S., and P. Duguid. 2000. The Social Life of Information. Boston: Harvard Business School Press.

Chu, W., and S. S. Keerthi. 2007. Support vector ordinal regression. Neural Computation. 19 (3): 792–815.

Duda, R. O., P. E. Hart, and D. G. Stork. 2001. Pattern Classification. 2nd ed. New York: Wiley.

Gross, P., A. Boulanger, M. Arias, D. L. Waltz, P. M. Long, C. Lawson, R. Anderson, et al. 2006. Predicting electricity distribution feeder failures using machine learning susceptibility analysis. AAAI. July. http://www.lsi.upc.edu/~marias/papers/iaai06.pdf

Hastie, T., R. Tibshirani, and J. H. Friedman. The Elements of Statistical Learning. New York: Springer.

Joachims, T. Learning rankings for information retrieval. http://videolectures.net/nips05_joachims_lrir/

Kearns, M. J., and U. V. Vazirani. 1994. An Introduction to Computational Learning Theory. Cambridge, Mass.: MIT Press.

Lin, C.-J. Support vector machines. http://videolectures.net/mlss06tw_lin_svm/

Machine Learning Summer School 2005—Chicago. http://videolectures.net/mlss05us_chicago/

Machine Learning Summer School 2006—Taipei. http://videolectures.net/mlss06tw_taipei/

Mitchell, T. 1997. Machine Learning. New York: McGraw-Hill.

———. 2006. The discipline of machine learning. http://www.cs.cmu.edu/%7Etom/pubs/MachineLearning.pdf

Murphy, C., G. Kaiser, and M. Arias. 2007. An approach to software testing of machine learning applications. SEKE. July. http://www.lsi.upc.edu/~marias/papers/seke07.pdf

Rouse, W. B. 1977. Human-computer interaction in multitask situations. IEEE Transactions on Systems, Man, and Cybernetics. SMC-7: 293–300.

———. 1988. Adaptive aiding for human/computer control. Human Factors. 30: 431–443.

Schölkopf, B., and A. J. Smola. 2002. Learning with Kernels: Support Vector Machines, Regularization, Optimization, and Beyond. Cambridge, Mass.: MIT Press.

Shapiro, R. Boosting. http://videolectures.net/mlss05us_schapire_b/

Shawe-Taylor, J., and N. Cristianini. 2000. An Introduction to Support Vector Machines and Other Kernel-Based Learning Methods. New York: Cambridge University Press.

Sutton, R. S., and A. G. Barto. 1998. Reinforcement Learning: An Introduction. Cambridge, Mass.: MIT Press.

von Krogh, G., K. Ichijo, and I. Nonaka. 2000. Enabling Knowledge Creation. New York: Oxford University Press.

Matrix

Akao, Y., ed. 1990. Quality Function Deployment: Integrating Customer Requirements into Product Design. Cambridge Mass.: Productivity Press.

Griffin, A., and J. R. Hauser. 1993. The voice of the customer. Marketing Science. 12 (1): 1–27.

Hales, R., D. Lyman, and R. Norman. 1994. QFD and the expanded house of quality. Quality Digest. February.

Hauser, J., and D. Clausing. 1988. The house of quality. Harvard Business Review. 32 (5): 63–73.

Karsak, E. E., S. Sozer, and S. E. Alptekin. 2002. Product planning in quality function deployment using a combined analytic network process and goal programming approach. Computers & Industrial Engineering. 44: 171–190.

King, B. 1989. Better designs in half the time: Implementing quality function deployment in America. Methuen, Mass.: GOAL/QPC.

Lowe, A., and K. Ridgway. 2000. UK user's guide to quality function deployment. Engineering Management. 10 (3): 147–155.

Mazur, G., and A. Bolt. 1999. Jurassic QFD. Paper presented at the 11th Symposium on Quality Function Deployment, Novi, Mich.

Pugh, S. 1991. Total Design: Integrated Methods for Successful Product Engineering. Wokingham, England: Addison-Wesley.

———. 1996. Creating Innovative Products Using Total Design: The Living Legacy of Stuart Pugh. Edited by D. Clausing and R. Andrade. Reading, Mass.: Addison-Wesley.

Werbos, P. J. 1974. Beyond Regression: New Tools for Prediction and Analysis in the Behavioral Sciences. Harvard University.

———. 1988. Generalization of backpropagation with application to a recurrent gas market model. Neural Networks. 1: 339–365.

———. 1989. Maximizing long-term gas industry profits in two minutes using neural network methods. IEEE Transactions on Systems, Man, and Cybernetics. 19 (2): 315–333.

———. 1994. The Roots of Backpropagation: From Ordered Derivatives to Neural Networks and Political Forecasting. New York: Wiley.

Business process modeling

Armenise, P., S. Bandinelli, C. Ghezzi, and A. Morzenti. 1993. A survey and assessment of software process representation formalisms. International Journal of Software Engineering and Knowledge Engineering. 3 (3).

Business Modeling FAQ.

Curtis, B., M. Kellner, and J. Over. 1992. Process modeling. Communications of ACM. 35: 75–90.

Dowson, M. 1988. Iteration in the software process. In Proceedings of the 9th International Conference on Software Engineering.

Feiler, P. H., and W. S. Humphrey. 1993. Software process development and enactment: Concepts and definitions. In Proceedings 2nd International Conference on Software Process.

Fernström, C., and L. Ohlsson. 1991. Integration needs in process enacted environments. In Proceedings of the 1st International Conference on the Software Process. IEEE Press.

Finkelstein, A., J. Kramer, and B. Nuseibeh. 1994. Software Process Modelling and Technology. New York: Wiley.

Harmsen, A. F., J. N. Brinkkemper, and J. L. H. Oei. 1994. Situational method engineering for information systems project approaches.

Rolland, C. 1993. Modeling the requirements engineering process. Paper presented at the 3rd European-Japanese Seminar on Information Modelling and Knowledge Bases.

———. 1997. A primer for method engineering. Proceedings of the INFORSID Conference.

Rolland, C., C. Pernici, and B. Thanos. 1998. A comprehensive view of process engineering. In Proceedings of the 10th International Conference, CAiSE'98, Lecture Notes in Computer Science 1413. Pisa, Italy: Springer.

Rolland, C., N. Prakash, and A. Benjamen. 1999. A multi-model view of process modelling. Requirements Engineering. 4 (4).

The smart grid

Deffeyes, K. 2001. Hubbert's Peak: The Impending World Oil Shortage. Princeton, N.J.: Princeton University Press.

Hoffert, M. I., K. Caldeira, G. Benford, D. R. Criswell, C. Green, H. Herzog, A. K. Jain, et al. 2002. Advanced technology paths to global climate stability: Energy for a greenhouse planet Science. 298: 981–987.

Lovins, A. 2003. Twenty hydrogen myths. http://www.rmi.org/

Simmons, M. 2003. World Oil. 6 (3): 76–80.

State of the Future. The Millenium Project. http://www.stateofthefuture.org/

Yergin, D. 1991. The Prize: The Epic Quest for Oil, Money, and Power. New York: Free Press.

INDEX

A

B

C

D

E

F

G

H

I

J

K

L

M

N

O

P

Q

R

S

T

U

V

W

ABOUT THE AUTHORS

Roger N. Anderson is Doherty Senior Scholar at the Lamont Doherty Earth Observatory and an adjunct professor in the Department of Earth and Environmental Sciences at Columbia University. Dr. Anderson is principal investigator of a team of scientists and graduate students in computer sciences developing Machine Learning software for intelligent control of the future electric grid of New York City. At Lamont, he founded and directed the Ocean Drilling Program Borehole Research, Global Basins, 4D Seismic Reservoir Simulation, and Energy Research Groups. Anderson received his Ph.D. from the Scripps Institution of Oceanography, University of California at San Diego, and his M.S. and B.S. from the University of Oklahoma in Geophysics. He holds 10 patents and has written three books and more than 200 peer-reviewed scientific papers.

Albert Boulanger is co-founder and Chairman of the Board of CALM Energy, Inc. and a member of the board of Team World Corps, a not-for-profit environmental and social organization. He is currently a Senior Staff Associate at Columbia University's Center for Computational Learning Systems. He held the CTO position of vPatch Technologies, Inc., a start-up company commercializing a computational approach to efficient production of oil from reservoirs. Before joining Columbia, he was at the Lamont-Doherty Earth Observatory and had also spent 12 years doing contract R&D at Bolt, Beranek, and Newman (now BBN Technologies). Boulanger started his career working for a cloud-seeding group at the National Oceanographic and Atmospheric Administration (NOAA) during high school and college at the University of Florida.

John A. Johnson is co-founder and CEO of CALM Energy, Inc. He has 20 years of experience in electric utility operations, engineering, strategy, and mergers and acquisitions. He started his career as an electrical engineer and has worked in most areas of an electric utility, including power plant operations and maintenance, nuclear plant technical services, transmission and substation technical services, controls engineering, electric distribution operations, electric system operations, energy management, and strategic planning. He received a Bachelor of Engineering from SUNY Maritime College and an MBA in finance from the Stern School of Business at New York University. He is a registered professional engineer in New York State.

Arthur Kressner is the Director of Research and Development, Power Supply, at the Consolidated Edison Company of New York, Inc., where he is responsible for developing and managing R&D at Con Edison and Orange & Rockland Utilities. Among his responsibilities during his career at Con Edison have been Chief Chemical Engineer and Plant Manager of the Arthur Kill and Ravenswood Power Generating Stations. Kressner has been in leadership roles at local, state and national levels in areas of energy policy, energy efficiency, customer end-use, and various national environmental and public health programs. Kressner has published papers and articles in peer-reviewed technical journals and industry magazines related to energy efficiency, information technology, and engineering models. He holds a bachelor's degree from Brooklyn Polytechnic and master's degree from New York University.